WEIRD SCIENCE
and BIZARRE BELIEFS

Gregory L Reece is an independent writer and scholar based in Montevallo, Alabama, whose research and publishing interests include philosophy of religion and the study of new religious movements. His previous books were *Elvis Religion: The Cult of the King* (2006) and *UFO Religion: Inside Flying Saucer Cults and Culture* (2007), both published by I.B.Tauris.

WEIRD SCIENCE

AND BIZARRE BELIEFS

MYSTERIOUS CREATURES, LOST WORLDS, AND AMAZING INVENTIONS

Gregory L. Reece

I.B. TAURIS

LONDON · NEW YORK

Published in 2009 by I.B.Tauris & Co Ltd
6 Salem Road, London W2 4BU
175 Fifth Avenue, New York NY 10010
www.ibtauris.com

In the United States of America and Canada
distributed by Palgrave Macmillan, a division of St Martin's Press
175 Fifth Avenue, New York NY 10010

ISBN 978 1 84511 756 6

A full CIP record for this book is available from the British Library
A full CIP record is available from the Library of Congress

Library of Congress Catalog Card Number: available

Typeset by JCS Publishing Services Ltd, www.jcs-publishing.co.uk
Printed and bound in the USA

This book is for Kristen.

Contents

Illustrations

Acknowledgments

Thanks to Tom Biscardi and all the members of the Searching for Bigfoot team, for an introduction to Bigfoot hunting in the wilds of Texas; to Craig Woolheater, for an enlightening conversation about Bigfoot; to Smokey Crabtree, for opening the box; to Dewayne Agin and the members of the Little Rock Grotto, for leading me safely to the center of the earth and out again; to the members of the Shaver Mystery chat group, for freely sharing their information and expertise; to my editor, Alex Wright, for patience and confidence; to Sandy, Cheryl, and Michael, for the flask and other things; to Sam and Olivia, for believing; and to Kristen, my one true love, for everything.

Introduction

It is 1977 and I am 10 years old and a reader of fantastic fiction, a reader of Wells, Verne, Burroughs, and Howard. I live by the exploits of Tarzan and John Carter of Mars, Professor Challenger and Phileas Fogg. I believe in the undersea kingdom of Atlantis and in Pellucidar, that strange land at the earth's core. Monsters walk the earth and soar through the heavens: the Wendigo, the banth, and the orluk prowl the earth for their prey; the pterodactyl, the malagor, and the sith hunt from the clouds.

My cousin Barry and I are playing in an abandoned corn crib. The air is dusty and hot. As we leave the building, the weathered gray boards of the door slam shut behind us. Looking up we see a terror – a roc, a malagor, an eagle the size of an airplane, sailing overhead, approaching lower and lower, as if intent on carrying one of us away to its mountain-top nest on the other side of the world.

We run for our lives and huddle together in the shelter of a tree. The creature passes on, its hunger unabated.

This is not a dream. It is real.

It is as real as the wasps that nest in corners and sting our hands and faces; as real as the redbud and mimosa trees that we climb in like squirrels; as real as the red dirt, black coal, and white limestone that is ever under our feet.

It is as real as the Bigfoot that haunts the Big Creek bottoms; as real as the root cellar that leads to secret underground kingdoms; as real as time travel and anti-gravity belts and laser beams.

It is as real as Jesus, as real as the Lord God Almighty.

I live in a world where time travel is as likely as air travel for a boy who has done neither; where werewolves are as likely as elephants for a boy who has seen neither; where Shangri La, or heaven itself, is as likely as New York or London for a boy who has traveled nowhere and seen nothing.

I am ten years old and mysterious creatures, lost worlds, and amazing inventions have not yet been removed from my world, harvested like trees from a pine forest. That will come later, as logic and science crowd out

fantasy and faith and as philosophy replaces theology. Adam and Eve will, ironically, go the way of the dinosaurs. Fact and fiction will no longer play together, side by side, like friends during the long summer vacation. The summer of childhood will come to its end. It will be time to go back to school (and to university and to graduate school). The malagor will be only a distant memory, clouded by time, overshadowed by more important things. It will come to seem like no more than a dream.
    But it was *real.*

*   *   *

Encounters with creatures as mysterious as the malagor are hardly rare. Accounts of giant birds of prey, large North American primates, and aquatic dinosaurs dwelling in the depths of Scottish lochs are almost a dime a dozen. Ask around. Everyone has a story or knows someone who has a story. Everyone has read a book or watched a television program about the Mothman of Virginia or the Manwolf of Wisconsin. The Six Million Dollar Man battled Bigfoot (a robot Bigfoot from outer space, no less) way back in the 1970s, a decade that also saw the creature team up for Saturday morning adventures with a young human protégé on *Bigfoot and Wildboy*. Mulder and Scully tracked Chupacabras through the 1990s. *The Water Horse*, Hollywood's version of the Loch Ness Monster, swam into theaters in the winter of 2007.

Of course this is to mix fiction and reality a bit. The space robot Bigfoot that battled Steve Austin is one thing, a fictional creation not much different from a Wookie. But there are people who actually claim to have seen Bigfoot, people who argue passionately that creatures such as this are real inhabitants of the American woodlands, people who say that there really is a plesiosaur dwelling in the waters of Loch Ness. There are people who go into the woods to hunt for Bigfoot equipped with guns and nets. There are people who seek scientific evidence for the existence of the Loch Ness Monster. What for many people are merely creatures of the imagination are, for others, creatures that really lurk in the darkness and the depths.

❝ THERE ARE PEOPLE WHO GO INTO THE WOODS TO HUNT FOR BIGFOOT EQUIPPED WITH GUNS AND NETS ❞

This blurring of fact and fantasy is not limited to the world of mysterious creatures. Everyone knows the story of Atlantis, the great civilization lost at sea. Captain Nemo and the crew of the *Nautilus* explored its undersea ruins. It is the home of Aquaman in DC Comics, and the home of his angrier cousin Namor, the Sub-Mariner in the Marvel Comics universe. Before Patrick Duffy played Bobby Ewing in that great 1970s soap opera *Dallas*, he was the Man from Atlantis for one brief season, donning fake gills and webbed hands to make the illusion complete. In the same decade, bell-bottomed adventurers explored Atlantium, the last city of Atlantis, while trying to escape the Bermuda Triangle in *The Fantastic Journey*. In *Stargate Atlantis*, the technology of the ancient Atlanteans is used to explore the wonders of the Pegasus galaxy.

We also know about the lost civilizations that are said to dwell beneath the surface of the earth. Its inhabitants have attacked Superman and the Fantastic Four. In the film version of Edgar Rice Burrough's *At the Earth's Core*, Doug McClure and Peter Cushing piloted a huge boring machine, the Iron Mole, deep underground and freed the primitive people of Pellucidar from the evil flying reptiles known as the mahars. They also saved the scantily clad Dian the Beautiful along the way. These are basic elements of science fiction and fantasy, of comic books and low-budget movies and television shows. But for some they are real. For some, Atlantis was a real place that continues to influence the development of our world. For some, the earth is hollow and the home of either an advanced race of enlightened beings or dangerously depraved creatures filled with wickedness and evil.

Likewise with the amazing inventions that make science fiction so interesting. You know, the time machines, ray guns, and anti-gravity devices that make the universes of *Star Trek* or *Star Wars* so much more exciting than our own. Luke Skywalker glided over the surface of Tatooine in a land-speeder. Michael J. Fox traveled backward and forward through time in a customized DeLorean. David Bowie, as real-life genius Nikola Tesla, invented a teleportation device in the film version of Christopher Priest's novel, *The Prestige*. Every sci-fi hero carries a ray gun, or phaser, or blaster. When science and technology are already doing so much for us, it is easy to imagine that they can also provide us with these things as well: a rocket ship to explore the universe, a time machine to travel into both the future and the past, and a laser gun to protect ourselves along the way. But, once again, what for most people are only dreams of the future are, for others, the facts of yesterday and today. The civilization of ancient India flew aircraft and controlled nuclear power. Time travel is possible today.

This book recounts stories of the mysterious creatures that prowl the hinterland of our inhabited spaces or that pop up unexpectedly in our suburban backyards. It recounts stories of the mysterious creatures that lumber out of the woods and into the headlights of our cars on dark deserted highways. It recounts stories of Bigfoot, the tall stinky man-ape; Bigfoot, the wild man of the woods; Bigfoot, star traveler and psychic guide. It recounts stories of Abominable Snowmen; of Puerto Rican goat-sucking Chupacabras; Wolfmen and Skinwalkers; Mothmen and Ropen; sea monsters from the depths. It is about horrors and dream creatures.

This book is about the lost worlds that call to us from the distant past, from the depths of the ocean, from the unfathomable dark of caverns and caves. It tells of Atlantis, mother of us all; Atlantis, city lost at sea; Atlantis, land of spirit and invention. It tells of Lemuria and Mu, Shambala and Shangri La. It tells of the world of the hollow earth; of Nazis, giants, mastodons, and stim rays. It tells of nightmare worlds and promised lands.

This book is about the amazing inventions and life-changing wonders that are produced, not in the R&D departments of multi-national corporations, but by precocious civilizations of the ancient past or by mad scientists ignored by Silicon Valley and Wall Street. It is about the technological wonders that promise to change our world. It is about the power of pyramids; about ancient calendars and the end of days; about flying chariots and nuclear bombs. It is about free energy and anti-gravity machines; about time travel and ray guns. It is about the wonders of the past and the marvels of the future.

This book is about mysterious creatures, lost worlds, and amazing inventions, and about the people for whom they are real.

*   *   *

In Part One we join a group of Bigfoot hunters and venture into the forests of north Texas in search of the mysterious creatures whose existence is known only through eyewitness testimony and muddy footprints, neither of which can stand long against time and the elements, slipping away into the far less convincing media of folklore and plaster casts. In Chapter One we examine the most famous of these creatures – Bigfoot. In Chapter Two we widen the scope to take in a whole world of creatures that lie just beyond the borders of accepted reality – creatures that won't be seen in any zoo. Chapter Three returns once again to the big guy, Bigfoot, though with an eye toward his more sensitive side. While there are plenty of people who view Bigfoot as nothing more than an as-yet-unrecognized animal, there are plenty of others who see the creature as something more, as a human-like being capable of language and art, or as a spiritual guide from another dimension or from beyond the stars.

In Part Two we follow wild stories, stories about blue-skinned people and Bigfoot armed with death rays, stories that lead us deep into underground caverns in search of lost worlds. Chapter Four takes us in search of the submerged societies of Atlantis, Lemuria, and Mu. In Chapter Five we tunnel even deeper into underground beliefs and borderland sciences and examine the tradition of subterranean civilizations and kingdoms, battling mystic Nazis and giant lizards along the way.

Part Three introduces consciousness-raising psychic brainwaves from outer space, first by looking to the forgotten technology of the past and then by looking around us to the inventions that hide on the margins of today's world. In Chapter Six we become entranced by the wonders of ancient India, Egypt, and America. We ride in the ancient Indian airships known as vimaanas, feel the power of the Egyptian pyramids, and tremble at the prediction of the end of the world found in the numerology of ancient Mayan calendars. In Chapter Seven we are exposed to today's fringe technology, mad scientists, and amazing inventions. We travel through time and space. We overcome the bonds of gravity. We draw power from the universe itself.

*   *   *

It is a challenge not to let my skepticism get the better of me, as an investigator of such weird and bizarre matters. It is too easy to laugh or to point out errors in logic when we confront beliefs that are remarkably different from what is taught in university classrooms. In this book I will do my best to avoid the easy path. This is not a book about critical thinking, nor is it a book about logical fallacies or bad science. It is a book about the diversity of human thought and about the plurality of beliefs that flourish whenever people are allowed to think freely. This book is not a skeptical look at Bigfoot, Atlantis, and pyramid power. It is, rather, an appreciative look.

This is not to imply that I believe any or all of the ideas presented here. My own worldview tends toward the tame and mundane. My science is rather pedestrian, my beliefs rather conventional. I do have a wild fascination with the ideas of other people, however, especially ideas that lead my brain through twists and turns it is not accustomed to taking. I like the people who, whether they know it or not, dare to be different, dare to think in an unapproved fashion about things that we are not supposed to think about once we are all grown up. I may be a rather boring person intellectually, but I get my kicks from those who take risks and are not afraid to make fools of themselves. In other words, I did not write this book as part of a strategy to try and stamp out weird science and bizarre beliefs. I wrote this book to *celebrate* weird science and bizarre beliefs. I believe that, in the long run, the world is better served by intellectual pluralism than by staid orthodoxy. (I will confess that I sometimes do offer critical appraisals of what I find out there in the borderlands. This happens, more often than not, at the times when I myself struggle with understanding. Some beliefs are so outlandish that my only response is to scratch my head. This puzzlement may sometimes come across as a predisposition to critique, which I hope it is not.)

I am also something of a pluralist when it comes to ontology, that is, the question of what is real and what is not. Traditional philosophical approaches usually divide up into the monistic and the dualistic sorts. Monists believe that there is only one kind of real thing, usually the material world but sometimes the world of spirit or mind. Dualists believe that there are two kinds of real things – the material and the spiritual. Modern science tends to be materialistically monistic, and traditional Christianity tends to be dualistic. I am a bit more unfettered when it comes to questions of what is real and what is not. I believe that there are all sorts of real things and that it gets us nowhere when we try to force things into one or the other of the aforementioned categories. Things can be real in different ways. We can only see that if we take the time to enjoy the diversity, if we bother to save our judgments until we have come to understand what people are saying and why. In this book I have tried to allow the reality of some of the weird scientific claims and bizarre beliefs to show through. Not that I am trying to convince the reader to believe in them. I am only trying to help the reader to appreciate them for what they are.

There are truly wonderful ideas out there in the world, a lot of truly wonderful people. There are people who believe that dinosaurs still roam the earth, people who believe that the earth is hollow, people who believe that the solution to our energy problems lies in the cosmic power that pervades the universe around us. I, for one, am glad they are there.

PART ONE

Bigfoot in Paris

ONE

Bigfoot

I drive into Paris, Texas wondering if this Paris has its own Eiffel Tower and
if cowboy hats, atop heads bronzed by the Texas sun, are here worn at jaunty
angles, like berets with brims. Is there a Texas Louvre with its own Venus de
Milo? Does this Paris, too, boast la Grande Cuisine?

Ouais.

A replica of the Eiffel Tower stands next door to the town's Civic Center,
its 65-foot frame looking like an oil rig built by a particularly artistic oilman,
a sensitive cowboy who dreamed of life on the River Seine rather than the Rio
Grande. At its zenith, at just the jaunty angle I had imagined, is a giant red
cowboy hat. Everything is not bigger in Texas, this Eiffel Tower is quite a bit
smaller than the real one, but the hat makes up for any deficiencies in size. (I
don't suppose Texans would want to make too much of that.) Paris, Texas,
alas, has no Louvre, though it does boast a cemetery monument depicting
Jesus in cowboy boots, the Texas answer to armless pagan goddesses. The
cuisine? La Grande it is – especially the chicken-fried steak.

It is none of these things that has brought me two hours north of Dallas,
however. Paris is only providing a motel room and a few quick meals. It is
but the jumping-off point for what promises to be a wild adventure. France
may have once been the home of Neanderthals and Paleolithic cave artists,
but the wilds of Texas offer the promise of living, breathing specimens of
hairy, bipedal hominids, wild men of the woods, throwbacks, and missing
links. I am here to join an expedition intent on capturing one. I am here to
hunt Bigfoot.

*　　*　　*

HAIRY GIANTS, WILD MEN, AND GORILLAS, OH MY!

The term 'Bigfoot' did not enter the popular lexicon until 1958. Bigfoot hunters and Bigfoot believers are quick to point out that this fact does not mean that the creatures in question were not known in North America long before that, however. Indeed, the popular press of the nineteenth and early twentieth centuries was replete with stories of hirsute giants, wild men, and gorillas that seem to fit the description of what would later be called 'Bigfoot.'

In his book *The Historical Bigfoot*, Chad Arment has tracked down many newspaper accounts of large, hairy, hominids that appeared at the end of the nineteenth and beginning of the twentieth centuries. For example, one report from the Newark, Ohio *Daily Advocate* of August 1, 1883 describes an encounter with such a creature in Ottawa, Ontario:

MAN OR GORILLA?
THE EXTRAORDINARY CHARACTER WHO IS SCARING CANUCKS

Ottawa, Ont., Aug. 1 – Pembroke, about one hundred miles north of Ottawa, has a lively sensation in the shape of a wild man eight feet high and covered with hair. His haunts are on Prettis Island, a short distance from the town, and the people are so terrified that no one has dared to venture on this island for several weeks. Two raftsmen named Toughey and Sallman, armed with weapons, plucked up sufficient courage to scour the woods in hopes of seeing the monster. About 3 o'clock in the afternoon their curiosity was rewarded. He emerged from a thicket having in one hand a tomahawk made of stone and in the other a bludgeon. His appearance struck such terror to the hearts of the raftsmen that they made tracks for the boat which was moored by the beach. The giant followed them, uttering demoniacal yells and gesticulating wildly . . . It is more than probable that the townspeople will arrange an expedition to capture, if possible, what Toughey describes as a man who looks like a gorilla, wandering about in a perfectly nude condition, and, with the exception of the face, completely covered with a thick growth of black hair. (160–1)

Arment also records this story from the Fort Wayne, Indiana *News* of November 18, 1908:

Wild Man of the Mountains
Robs Traps of Prey and Commissary of Supplies
Peculiar Footprint Left by the Man-Animal as He Flees From His Would-Be Captor

Special Correspondence

Santa Monica – Nov. 18. – Is there a wild man roaming the mountains of the Santa Monica range? There is, according to the story told by Bertrand Basey, who has just come down from the vicinity of Point Dume, where he has been acting as commissary for the contractors who are engaged in the construction of the Malibu-Rindge railway.

Basey says that there were frequent losses from the improvised store house . . . He kept a strict lookout and was soon rewarded by the sight of a brown being fashioned after the form of a man, approaching the tent. The thing was on all fours, was devoid of clothing or such covering as might have been provided by the skins of animals, and had a face covered with hair . . . Suddenly the thing gave forth a guttural yell, rose upright on its hind legs and disappeared in the underbrush.

> ❝ SUDDENLY THE THING GAVE FORTH A GUTTURAL YELL, ROSE UPRIGHT ON ITS HIND LEGS AND DISAPPEARED IN THE UNDERBRUSH ❞

Nothing more was seen or heard of the mysterious half-human beast, although the railroaders who went to the beach for a bath that morning are still wondering what manner of animal had been there before them to leave peculiar tracks in the sand. The tracks which they photographed were not unlike those that would have been left by the hands and feet of a man were he provided with long claw-like nails for each of his five toes and the four fingers and thumb. (106)

The Lincoln, Nebraska *Star* even managed to work in a reference to Edgar Rice Burroughs' famous Tarzan character in their report of October 22, 1922:

Tarzan Comes to Life on Western Slope of Rockies

Denver, Colo., Oct. 21 – Weird stories of the ravages of a reported 'wild man, half man, half beast' in the Naturita valley in a thinly settled district on the western slope of the Colorado Rockies, reached Denver tonight.

According to the 'eye witness' accounts, the strange hairy creature leaps across the waste places on all fours and subsists on the raw flesh of fowls and animals. Its body is gaunt and covered with hair and the eyes gleam ferociously from under shaggy brows, residents of the valley who claim to have seen the animal, declare.

The reported hybrid has invaded barnyards, killed chickens, robbed hens' nests and fled on the appearance of any person, the accounts stated.

Ranchers of the Naturita district are said to be planning a hunt in an effort to capture the creature. (109)

A quick review of just these three accounts reveals a less than homogeneous picture of the kind of creature that is the subject of the encounters. In two accounts the animal crawls on all fours like a dog (and leaves footprints, a staple of modern Bigfoot accounts). In the other it stands upright and wields weapons, demonstrating some technological accomplishment. In one account the apparition is covered all over with hair, with the exception of its face. In another the creature 'had a face covered with hair.' Despite these differences the stories do share basic features, however. All three newspapers report eyewitness accounts (though the *Star* places the term 'eye witness' in quotation marks) of encounters with strange creatures that, though described variously as wild men, giants, gorillas, and half-man/half-beast hybrids, are nevertheless large, hairy, hominids.

THE ABOMINABLE SNOWMAN

The North American continent was not the only source for hairy hominid stories in the early twentieth century. Many of the most important accounts from the period came from the Himalayan Mountains. Ivan Sanderson's retelling of the Himalayan accounts from 1920 is usually regarded as the classic version. He wrote:

In that year an incident occurred that was impressive enough but which might have been either wholly or temporarily buried had it not been for a concatenation of almost piffling mistakes. In fact, without these mistakes it is almost certain that the whole matter would have remained in obscurity and might even now be considered in an entirely different light or in the status of such other mysteries as that of 'sea-monsters.' (10–11)

1. 1957's *Abominable Snowman of the Himalayas*:
'A shock-fest for your scare-endurance'.

According to Sanderson, the hairy hominid of the Himalayas first came to widespread public attention in the west following a telegram sent by Lt. Col. C.K. Howard-Bury concerning the results of a reconnaissance expedition in the region of Mount Everest. Approaching the northern face of Everest, at around 17,000 feet, the expedition witnessed a group of dark shapes moving about in a field of snow higher up the mountain. After considerable effort the expedition made their way to the field. The creatures were long gone, but several well-defined footprints were found in their place. These prints were three times larger than normal human prints. Howard-Bury assumed that they were the tracks of large wolves. The Sherpa guides, however, disagreed. They claimed that the tracks were from a human-like creature sometimes seen in the area. The name given by the Sherpas to this hominid seems to have been garbled, either by Howard-Bury or by Henry Newman, the recipient of the telegraph. Newman relayed the message to the world, translating the garbled name into English as 'Abominable Snowman.' A legend was born.

Paranormal researcher John A. Keel reports other examples of sightings from the 1920s and 1930s that served to reinforce the legend, including a 1924 expedition to Everest, led by General Bruce, that reported footprints and 'a great, hairy, naked man running across a snow field' at 17,000 feet (Keel 2002, 63). Visiting the area in 1955–6, Keel claims to have learned that the Sherpas accepted the reality of the phantom without question. They also greatly feared it, perhaps with good reason. According to Keel, a Sherpa herdsman named Lakpha Tensing was killed by one of the creatures in 1949 while in the Nanga Parbat pass. Fear of the creature was so great that mothers warned ill-behaved children that the monster will take them if they are not careful.

One interesting detail noted by Keel was the belief among the locals that the feet of the snowman are mounted backwards. This belief originates with an encounter with the beast from the early 1900s. According to the story, a group of local men employed by the British to install a telegraph line over a remote mountain pass never returned to their base. The next day British soldiers went in search of them. They did not find the missing men, but they did find a hairy, naked man-beast hiding in the crevice under some boulders. They shot the creature and hauled it to a small bungalow, where it was packed up and shipped back to England. Though the carcass was never seen again, one man that Keel interviewed claimed to have seen the remains of the animal when he was a small boy. He described the animal as being ten feet tall and covered with long shaggy hair. Its face was hairless, its mouth was filled with yellow teeth, and its eyes were red. Most striking of all, the man claimed that the creature's feet were attached backward, with the toes pointing to the rear. Keel speculates that the animal's feet may have been hand-like, like an ape's, and thus caused the confusion. (64–5)

At least two important expeditions would set out for Mount Everest with the express purpose of tracking the Snowman, soon called by another local name, the Yeti or 'mysterious creature.' In 1957, Texas oilman and self-proclaimed monster hunter, Tom Slick, explored the mountain with limited results. A few years later Sir Edmund Hillary would take on the challenge, producing a report of his findings in the October 1962 issue of *National Geographic*. In Hillary's assessment, Yeti tracks could be accounted for by the tendency of tracks in the snow to melt and elongate in the sun, and then refreeze when the weather turns colder. Such a process could transform something as mundane and uncontroversial as fox tracks into large, human-like prints. Hillary also examined what locals claimed to be the fur and scalp of a Yeti. The fur turned out to be from a Tibetan blue bear and the scalp from a serow, a goat-like animal. Hillary's negative conclusions did little to dampen the enthusiasm for the Abominable Snowman, however. Real creature or not, the name itself was certainly too dramatic ever to go away. Indeed, just two years after Hillary's report was published, 'Bumble,' 'The Abominable Snow Monster of the North,' was a featured character in the television adventures of Rudolph the Red-Nosed Reindeer.

The Yeti has also continued to be seen in the wild, as well as on television, despite Hillary's claims. Loren Coleman and Jerome Clark report that the American Craig Calonica encountered two of the beasts in 1998, on the Chinese side of Everest. Calonica described the creatures, who were walking together, as having thick black fur, long arms, and large hands. Italian Reinhold Messner asserts that he saw the Yeti on four separate occasions in 1986 and 1997. Messner's claims about the creature's identity have gone against the grain of mainstream Yeti theories, however. Instead of a large ape-like creature, Messner suggests that the Yeti may be an as-yet-undiscovered species of giant bear.

SASQUATCH

While the Abominable Snowman was haunting the Himalayas, school teacher J.W. Burns found accounts of hairy hominids throughout the Native American folklore of western Canada. In the 1920s Burns worked as a teacher on the Chehalis Indian Reservation, near Vancouver, British Columbia. Intrigued by stories told by the locals of hairy giants, he began to seek out and record the tales, both folkloric and eyewitness accounts. In April 1929 his article 'Introducing B.C.'s Hairy Giants' was published in *MacLean's* magazine, drawing widespread attention, for the first time, to the legend of the Sasquatch. His stories were subsequently told and retold in other magazines and newspapers.

Among those whose stories Burns tells is Peter Williams, a resident of the Chehalis Reserve. In the month of May, some twenty years earlier, Williams claimed to have encountered a Sasquatch a mile from the reserve. Hearing a nearby grunt, he turned to see what he thought at first was a huge bear, crouching on a rock. As Williams lifted his gun to shoot, the creature stood up on two legs and let out a scream. Instead of a bear, Williams beheld a man, over six feet tall and covered with hair. In a rage, the creature jumped to the ground and charged. Williams ran to the river and leaped into his boat, thinking that he had escaped. The creature, however, simply waded across the fast-moving water. Reaching his home, Williams rushed in and locked and barricaded the door.

'After an anxious waiting of twenty minutes,' resumed the Indian, 'I heard a noise approaching like the trampling of a horse. I looked through a crack in the old wall. It was the giant. Darkness had not yet set in and I had a good look at him. Except that he was covered with hair and twice the bulk of the average man, there was nothing to distinguish him from the rest of us. He pushed against the wall of the old house with such force that it shook back and forth. The old cedar shook and timbers creaked and groaned so much under the strain that I was afraid it would fall down and kill us. I whispered to the old woman to take the children under the bed.' (www.westcoast-sasquatch.com)

After a while the creature gave up his assault on the house and disappeared back into the wilderness. The next morning, Williams found tracks in the mud. They were 22 inches long.

Burns also includes the story of Charley Victor, of the Skwah Reserve. It is Victor who introduces the term 'Sasquatch,' or 'hairy mountain men,' to describe the giants. Victor claimed to have had many encounters with the creatures, including one encounter that resulted in the shooting of a Sasquatch youth. When his dog treed what he thought was a bear, Victor shot his gun only to learn that it was a naked boy, 12 or 14 years of age. The injured boy began to howl for help and was soon answered by a Sasquatch woman, covered with hair and standing around six feet tall.

'In my time,' said the old man, 'and this is no boast, I have in more than one emergency strangled bears with my hands, but I'm sure if that wild woman laid hands on me, she'd break every bone in my body. She cast a hasty glance at the boy. Her face took on a demoniacal expression when she saw he was bleeding. She turned upon me savagely, and in the Douglas tongue said: 'You have shot my friend.'

I explained in the same language – for I'm part Douglas myself – that I had mistaken the boy for a bear and that I was sorry. She did not reply but began a sort of wild frisk or dance around the boy, chanting in a loud voice for a minute or two, and, as if in answer to her, from the distant woods came the same sort of chanting troll. In her hand she carried something like a snake, about six feet in length, but thinking over the matter since, I believe it was the intestine of some animal. But whatever it was, she constantly struck the ground with it. She picked up the boy with one hairy hand, with as much ease as if he had been a wax doll.

While Victor felt certain that the female was of the Sasquatch tribe, he was also convinced that the boy was not. Not only was his skin white and mostly hairless, but the female had called him her friend. Victor surmised that the Sasquatch people must have stolen him.

Perhaps Burns' most tantalizing story is his account of the Native American girl Serephine Long, who claimed to have been kidnapped and forced to live among the Sasquatch. Burns records her story:

I was walking toward home one day many years ago carrying a big bundle of cedar roots and thinking of the young brave Qualac (Thunderbolt), I was soon to marry. Suddenly, at a place where the bush grew close and thick beside the trail, a long arm shot out and a big hairy hand was pressed over my mouth. Then I was suddenly lifted up into the arms of a young Sasquatch. I was terrified, fought, and struggled with all my might. In those days, I was strong. But it was no good, the wild man was as powerful as a young bear. Holding me easily under one arm, with his other hand he smeared tree gum over my eyes, sticking them shut so that I could not see where he was taking me. He then lifted me to his shoulder and started to run.

> ❝ AT A PLACE WHERE THE BUSH GREW CLOSE TO THE TRAIL, A LONG ARM SHOT OUT AND A BIG HAIRY HAND WAS PRESSED OVER MY MOUTH ❞

After a long journey in which the Sasquatch carried her across a river, up and down hills and mountains, and through a long tunnel, Serephine was placed on the ground.

I heard people talking in a strange tongue I could not understand. The young giant next wiped the sticky tree gum from my eyelids and I was able to look around me. I sat up and saw that I was in a great big cave. The floor was covered with animal skins, soft to touch and better preserved than we preserve them. A small fire in the middle of the floor gave all the light there was. As my eyes became accustomed to the gloom I saw that beside the young giant who had brought me to the cave there were two other wild people – a man and a woman. To me, a young girl, they seemed very, very old, but they were active and friendly and later I learned that they were the parents of the young Sasquatch who had stolen me. When they all came over to look at me I cried and asked them to let me go. They just smiled and shook their heads. From then on I was kept a close prisoner; not once would they let me go out of the cave. Always one of them stayed with me when the other two were away.

Serephine claimed to have lived with the giants for nearly a year. During that time she was fed roots, fish, and meat. She even learned a few words of their dialect, which was similar to the Douglas language. Finally, after becoming dangerously ill, she was able to convince the young Sasquatch to take her back to her home. She pleaded with him to allow her to see her family before she died. Sticking her eyelids together once again, the young male carried her back to her village. She described her return home to Burns:

> My people were all wildly excited when I stumbled back into the house for they had long ago given me up as dead. But I was too sick and weak to talk. I just managed to crawl into bed and that night I gave birth to a child. The little one lived only a few hours, for which I have always been thankful. I hope that never again shall I see a Sasquatch.

Burns' Native American Sasquatch stories were received with delight by residents of the area, and for a while the village of Harrison Hot Springs hosted an annual Sasquatch Festival. Though interest had largely waned by that time, in 1957 the village once again took advantage of the notoriety sparked by the tales to draw attention to their region during British Columbia's centennial celebration. In honor of the occasion the village council at Harrison Hot Springs decided to use funds designated for the celebration to sponsor a Sasquatch hunt. According to Bigfoot researcher John Green's account, the idea proved to be a great success. Though the hunt never actually took place, the story was picked up by the press and served to focus attention on the region and revive interest in the Sasquatch (Green, 4). Soon, sparked by the renewed interest in the creatures, individuals came forward with new stories of encounters with Sasquatch. These tales would take the basics of Burns' stories in new directions and would come to be regarded as classics in the field of Bigfoot studies.

THE CLASSICS

The first story to surface following the Sasquatch revival of 1957 was told by William Roe, a resident of Cloverdale, British Columbia. John Green, owner and editor of a local weekly newspaper, asked Roe to put his story in the form of a sworn affidavit. According to this sworn account, Roe's encounter with one of the creatures took place in October 1955. Working on a highway construction project that cut through the wilderness, Roe decided to take a break and explore the area, particularly an abandoned mine some five miles from the construction site. He caught sight of the mine at around three in the afternoon. As he walked into a clearing he caught sight of something else.

> I saw what I thought was a grizzly bear, in the brush on the other side . . . I could just see the top of the animal's head and the top of one shoulder. A moment later it raised up and stepped out into the opening. Then I saw that it was not a bear. (Green, 9)

The creature stood about six feet tall, was about three feet wide, and weighed around three hundred pounds. It was completely covered with dark brown, silver-tipped hair. As it approached Roe he could tell by its breasts that the creature was a female. Despite the presence of breasts, its body shape was 'not curved like a female's' but 'straight from shoulder to hip.' The creature's arms were much bigger than a human's and longer, reaching practically to the knees. The feet were broad, five inches at the front and narrowing toward the heels. 'When it walked it placed the heel of its foot down first, and I could see the gray-brown skin or hide on the soles of its feet.'

Roe's description continued:

> The head was higher at the back than at the front. The nose was broad and flat. The lips and chin protruded farther than its nose. But the hair that covered it, leaving bare only the parts of the face around the mouth, nose, and ears, made it resemble an animal as much as a human. None of its hair, even on the back of its head was longer than an inch, and that on its face was much shorter. Its ears were shaped like a human's ears. But its eyes were small and black, like a bear's. And its neck was unhuman. Thicker and shorter than any man's I had ever seen. (9–10)

Soon, the creature became aware of Roe's presence and began to walk back the way it had come. Roe wrote, 'For a moment it watched me over its shoulder as it went, not exactly afraid, but as though it wanted no contact with anything strange' (10).

Roe thought momentarily about shooting the creature with his rifle but decided against it, considering its human-like qualities. After the creature left he examined its scat and found no evidence of anything other than a vegetarian diet. He also found the place where it had bedded for the night. He found no indication of tools or technology.

Following the publication of this account in the local newspaper, Albert Ostman came forward with his story of a Sasquatch encounter. Unlike Roe's account, Ostman's took place many years before, in 1924, and seems reminiscent of the accounts published at that time by Burns. Ostman claimed that while dozing in his sleeping bag deep in the woods on a prospecting trip he was suddenly awakened by something lifting him in the air. Thinking at first that he might be in a snow slide or that someone had placed him on a horse, he slowly realized that he was being carried by something walking on two legs.

Unable to reach his knife and cut his way out of the sleeping bag, Ostman was forced to go along for the ride, clutching his rifle. After some while he could tell that he was traveling uphill. Whoever, or whatever, was carrying him began to breathe deeply and let out a slight cough. It was then that Ostman knew he had been captured by a Sasquatch. After a long journey, uphill and down, Ostman was dropped to the ground. He worked his way out of his sleeping bag and looked around:

> It was still dark, I could not see what my captors looked like. I tried to massage my legs to get some life in them, and get my shoes on. I could hear now it was at least four of them, they were standing around me, and continuously chattering. I had never heard of Sasquatch before the Indian told me about them. But I knew I was right among them ... I asked, 'What you fellows want with me?' Only some more chatter. It was getting lighter now, and I could see them quite clearly. I could make out forms of four people. Two big and two little ones. They were all covered with hair and no clothes on at all ... They looked like a family, old man, old lady and two young ones, a boy and a girl. The boy and the girl seemed to be scared of me. The old lady did not seem too pleased about what the old man dragged home. But the old man was waving his arms and telling them all what he had in mind. They all left me then. (Green, 14)

Fortunately, Ostman had plenty of food in a backpack he had stored in the bottom of his sleeping bag. He also had a knife, some matches, and rifle shells.

According to Ostman, the old female was very meek, the young male was downright friendly, and the young female seemed harmless.

Unlike Roe's female, this creature's chest was 'flat like a boy's.' Ostman believed that, given the chance, he could have transported the young female back to civilization, and then placed her in a cage for public display. As he was able to observe them up close, Ostman was able to offer a rather detailed description of the Sasquatch:

> The young fellow might have been between 11–18 years old and about seven feet tall and might weigh about 300 lbs. His chest would be 50–55 inches, his waist about 36–38 inches. He had wide jaws, narrow forehead, that slanted upward round at the back about four or five inches higher than the forehead. The hair on their heads was about six inches long. The hair on the rest of their body was short and thick in places.
>
> The women's hair on the forehead had an upward turn like some women have – they call it bangs, among women's hair-do's. The old lady could have been anything between 40–70 years old. She was over seven feet tall. She would be about 500–600 pounds. She had very wide hips, and a goose-like walk. She was not built for beauty or speed. Some of those lovable brassieres and uplifts would have been a great improvement on her looks and her figure.
>
> The man's eyeteeth were longer than the rest of the teeth, but not long enough to be called tusks. The old man must have been near eight feet tall. Big barrel chest and big hump on his back – powerful shoulders, his biceps on upper arm were enormous and tapered down to his elbows. His forearms were longer than common people have, but well proportioned. His hands were wide, the palm was long and broad, and hollow like a scoop. His fingers were short in proportion to the rest of his hand. His fingernails were like chisels. (16)

Ostman's escape from the creatures came as a result of the creatures' interest in his can of snuff. After watching Ostman place a large portion of the tobacco in his mouth, the 'old man' grabbed the box away from him and emptied its contents into his own mouth. He then licked the inside of the box with his tongue. The results were astonishing. In just a few minutes Ostman could see the 'old man's' eyes rolling over in his head. He was sick. The creature grabbed a can of cold coffee and drank it down, grounds and all. He then put his head between his legs and rolled away from Ostman, beginning to squeal. Ostman saw this as his chance, grabbed his rifle, and began to run. He fired one shot at the adult female who was in pursuit, turning her back. Ostman then made his escape, and after some difficulty found his way back to civilization.

2. The 'Searching for Bigfoot' equipment trailer:
'Here Come's [sic] the Bigfoot Hunter!'

*   *   *

As I arrive at my hotel room (cost included in the online ticket purchased from parapalooza.com that allows me to participate in this Sasquatch safari), I notice two men hovering around my door. I am tired from my flight and the drive up from Dallas and am hoping to take a short nap before the action is scheduled to start later in the evening, but these guys seem to have a different idea.

'Are you Greg?' one asks.

'Yeah.'

'Come on – we need to get going before it rains. Can we take your truck?'

'Sure, just give me a couple of minutes to get my boots on.'

Boots on, we are soon driving out of town, with Paris in our rearview mirror.

My companions are T.J. and Andy. T.J. (Tom Junior) is the son of Tom Biscardi, leader of the expedition, or as he says, 'the man who put the "Big" back into Bigfoot!' Andy is a paying customer, like me. He has come to learn the tricks of the Bigfoot trade, hoping to catch his own view of the creature, and to become a part of history if we manage to achieve our mission as T.J. describes it – capture a Bigfoot. We are equipped, T.J. informs us, with sensitive tracking equipment, tazers, and a cannon that will fire an electrified net. A helicopter is on standby to transport the creature to an undisclosed location, where it will be studied and cared for, before being released back into the wild. This is going to be big.

T.J. directs me to take a left, then a right, then another right. As the road turns to gravel I see a truck and trailer parked on the shoulder. The trailer reads 'Here Come's the Bigfoot Hunter.' Superfluous apostrophe or not, this must be the place. We climb out of the truck and begin to pick our way across the mud into the woods. I try to call my wife to let her know that I have made it safely to Paris and that I am about to walk into the woods with total strangers, but I have no cell phone signal.

There are four or five people standing around a clearing in the woods, including one with professional video equipment. I recognize Tom Biscardi among them. He calls us over to examine a set of tracks that have been found earlier in the day. They have been waiting for me to arrive before they start making plaster casts and are in a hurry because of the threat of rain. T.J. begins mixing plaster as the rest of us gather around the prints.

There are four of them in a line, humanoid in appearance, but much bigger than any human footprint. The cameraman moves in for a close up of the tracks, then of Tom describing the tracks. Tom shows us how the creature had stepped into the muddy clearing, taken a few steps, and then moved back under the trees and on to solid ground. He explains how the tracks show 'dermal ridges,' those wrinkles and lines in skin that indicate a track is legitimate and not the imprint of a wooden mold. Tom is noticeably impressed by the size of these tracks and by the large stride indicated by the distance between the prints.

'It's a big one,' he says, 'a big bull.'

BIGFOOT

On October 6, 1958 a small northern California newspaper ran a story entitled:

NEW 'SASQUATCH' FOUND
IT'S CALLED BIGFOOT

The story was accompanied by a photograph of Jerry Crew holding a plaster cast of a very large human-like footprint. Jerry's story began on August 27 of that same year. On the morning of that day Crew went to work as a heavy-equipment operator employed in the construction of a lumber road in the wilderness near the Oregon border. He was

❛ THE TRACKS LOOKED LIKE BARE HUMAN FEET, EXCEPT THAT THEY WERE 17 INCHES LONG ❜ employed by Ray Wallace, who was a subcontractor working for the National Parks Service. Crew's equipment had been parked overnight at the job site. Checking his equipment before going to work, Crew discovered tracks leading up to the machine, all around it, and then back into the woods. The tracks looked like bare human feet, except that they were 17 inches long. Upon further inspection, Crew discovered that the tracks led down a steep incline, with the stride extending up to 60 inches in some places. Crew showed the tracks to his co-workers, who were just as perplexed as he was. About a month later, the tracks reappeared.

The story of what the men had discovered began to be circulated throughout the small communities in the region. The *Humboldt Times* of Eureka, California printed a letter from the wife of one of the work crew, describing the prints. A few letters followed, each confirming what had been reported. On October 2 the tracks returned, for three nights in a row. Jerry Crew, perhaps wanting to establish that there was indeed more to the story than just his fertile imagination, made a plaster cast of one of the tracks. While on a trip to Eureka a couple of days later, Crew showed the cast to a friend. When the newspaper heard that he was in town they asked to take a photograph of him holding the cast. The wire service picked up the story – 'Bigfoot' had arrived on the scene.

Soon, other reports came in from the construction site. Ray Wallace's brother Wilbur reported that a 55-gallon drum of fuel had been moved and thrown down into a deep ravine. In addition, a steel culvert was found at the bottom of another bank, and a 700-pound tire had been rolled a great distance down the road and then hurled into a ravine. A photographer and reporter from the *Humboldt Times* had gone to the site to photograph a series of tracks and had found a human-like, but incredibly large, pile of scat.

The Bluff Creek Incident, as it has come to be called, marked a turning point in the story of hairy hominids in North America. While the Sasquatch stories reported by Burns, Roe, and Ostman, as well as the early newspaper accounts of hairy giants, apes, and wild men, tended to focus on eyewitness accounts, the Bigfoot experiences of northern California introduced a piece of empirical evidence that had not been seen before. While the Sasquatch stories could be regarded as merely folklore or tall tales, footprints were quite another matter.

Crew's plaster cast and the catchy new name, 'Bigfoot,' guaranteed that focus would now be placed on the tracks of the creature in question. Physical evidence trumped eyewitness accounts and ancient legends. This new focus on empirical evidence meant that Sasquatch/ Bigfoot would become a different kind of phenomenon. The study of folklore, after all – whether the study of the myths of indigenous peoples, newspaper accounts from days gone by, or eyewitness reports and tales of local inhabitants – is basically anthropology. It is the study of human beings and what human beings believe or once believed. The analysis of tracks is quite another matter. Sasquatch had resided in the land of leprechauns and brownies; Bigfoot lived in the world of forensic science.

The pioneer in the field of Bigfoot track analysis was the late Grover Krantz, former professor of anthropology at Washington State University and the first 'academic' researcher to study the Bigfoot/Sasquatch phenomenon as a biological, rather than a cultural, subject. The heir to Krantz's legacy is Jeff Meldrum, professor of anatomy and anthropology at Idaho State University. Meldrum's non-Bigfoot research is in the field of evolutionary biology and primate locomotion, making his interest in the hairy hominids of North America and their mysterious tracks quite understandable.

Meldrum's magnum opus is called *Sasquatch: Legend Meets Science*, a companion to a documentary film of the same name. In this work, Meldrum seeks to gather together the best empirical evidence for the existence of Bigfoot. One of his most critical sections is an in-depth look at the footprint evidence. Meldrum begins:

> Clearly, *something* is leaving enormous humanlike tracks on the backcountry roads and riverbanks of North America's mountain forests. With the potential for misidentification of bear or human footprints accounted for, the otherwise inexplicable footprints that remain must be either hoaxed or hominid. In the absence of bones or body, the tracks constitute the most abundant and informative data that can be dealt with by scientific evaluation. (221)

He describes the footprint evidence this way:

> The sometimes-enormous size of the sasquatch tracks gave rise to the common American appellation of 'Bigfoot.' These footprints average between 15 and 16 inches in length, with a reported range of 4–27 inches. Their superficially humanlike appearance is largely the consequence of the inner big toe being aligned with the remaining toes, whereas an ape's inner toe diverges much like a thumb. The

resemblance to human footprints largely stops there, however. In fact, the sasquatch footprints lack the principal distinctive features that set the human foot apart from its hominoid cousins. Sasquatch footprints are typically flat with no consistent indication of the true hallmark of the human foot – a fixed longitudinal arch. Additionally, there is little indication of differential weight bearing under a specialized 'ball' at the base of the big toe. The sasquatch foot is relatively broader and the sole pad apparently thicker, by comparison to human feet. The heel and toe segments are disproportionately longer. (223)

Sasquatch prints usually have an extremely long stride. The footprints are often found one directly in front of the other, unlike human prints that usually alternate to the left and right of center. Human footprints usually show a variation of depths, owing to the arched nature of the foot, while Sasquatch prints tend to be uniform in depth. Meldrum writes:

In all, the sasquatch footprint is not merely an enlarged facsimile of a human footprint, but appears to represent a uniquely adapted primate foot associated with a distinctive mode of bipedalism, one that may well have evolved independently through roughly a parallel to hominid bipedalism. (224)

And in addition:

The inferred architecture of the sasquatch foot is not only well documented, but seems well suited to the physical aspects of the terrain of its purported range. The retention of somewhat prehensile toes, combined with increased leverage of the heel, give it an advantage negotiating the steep and uneven mountainous forest landscapes of North America. The locomotor adaptation of an organism is a major element in defining its niche. The conformity of the inferred sasquatch locomotion to an overall hominoid/early hominid framework, and the anatomical distinctions correlated to its environmental specializations are plausible and compelling arguments for a real animal. (247–8)

In addition to his study of the morphology of the Bigfoot foot, Meldrum also focuses a good deal of attention on dermatoglyphics, the study of skin features, which includes the lines and ridges that are present on primate feet. Meldrum identifies Officer Jimmy Chilcutt as one of the leading experts in the study of these dermal ridges. Chilcutt is a crime scene investigator and latent fingerprint examiner employed by the police department of Conroe, Texas.

He also has extensive experience fingerprinting primates at zoos and research centers. He has developed quite a reputation in the Bigfoot community as an expert in the identification of legitimate tracks. According to Chilcutt, the dermal ridges, rather than the shape and size of a track, are the most important indication of authenticity. After all, Sasquatch feet, like those of humans, come in all shapes and sizes. They also can look an awful lot like bear tracks. But bear paws, human feet, and wooden hoaxer shoes do not have the distinctive Sasquatch dermal ridges, which are wider than those of humans and run the length of the foot. The surest scientific method for identifying Bigfoot tracks is thus to examine the lines and swirls left imprinted in the mud. Indeed, Meldrum argues:

6 AFTER ALL, SASQUATCH FEET, LIKE THOSE OF HUMANS, COME IN ALL SHAPES AND SIZES 9

> If repeated independent occurrences of dermatoglyphics in sasquatch footprint casts spanning several decades, with hundred of miles of geographic separation, and displaying consistent yet distinct features of ridge texture and details of flow pattern can be confirmed, it would constitute compelling evidence for an unknown primate. (259)

As well as footprints, Meldrum also cites an important discovery, made in September 2000, of a partial body imprint of a Sasquatch. The imprint was found at a location known as Skookum Meadows in the Cascade Mountains of southern Washington by the Bigfoot Field Researchers Organization (BFRO). Placing a pheromone attractant and several pieces of fruit near a muddy turnout adjacent to a Forest Service road, the BFRO was rewarded the next morning with intriguing evidence. The pheromone attractant had been removed from the tree and most of the fruit was missing or partially eaten. Most important, however, was an imprint of some sort of creature found in a muddy depression. Meldrum describes what was found:

> The impressions appeared to include that of a left forearm, buttocks, thigh, and heels ... Apparently the sasquatch had approached the puddle, lain down across the halo of moist soil on the periphery of the puddle without stepping there, then leaned onto its left elbow and forearm to reach in with its right arm toward the puddle for a sampling of the fruit, while pushing against the mud with it heel. (115)

Upon close examination of a plaster cast made from the imprint, Meldrum found what he believes to be important evidence:

> Of particular interest to me was what could only be interpreted as a distinct heel impression. As I meticulously removed the encrusting soil, it appeared that the heel bore skin ridge detail. Once the heel was thoroughly cleaned, a thin latex peel was made of the skin detail. Consultations over the apparent dermatoglyphics, or skin ridges, were had with latent fingerprint examiner, Officer Jimmy Chilcutt. He found them to be consistent in texture and appearance with other specimens of purported sasquatch tracks exhibiting such skin ridge detail. (117)

Meldrum, like Grover Krantz before him, believes that such empirical evidence is critical for the study of North American hairy hominids. Recognizing that eyewitness testimony will always be insufficient, physical evidence must be discovered and analyzed. In the world of Sasquatch research, legend must be corroborated by science. Footprints, constituting the largest body of evidence for the creature, must be subjected to scientific scrutiny and analysis. As Meldrum notes, once we rule out cases of mistaken identification – bear or human tracks – we are left with only two choices: hoaxes or hominids. The question then is one of proving that the Bigfoot prints are real and not hoaxes. It is in defense of the legitimacy of the tracks that Meldrum marshals his evidence. The latter half of the twentieth century was a time of plaster casting and track analysis in the Bigfoot world – from Jerry Crew's footprint cast from the Ray Wallace construction site, to the Skookum body cast. Bigfoot believers proclaimed the tracks to be real, skeptics proclaimed them hoaxes. Then in 2003 we heard once again from Ray Wallace.

Upon his death in November 2002, Ray's family proclaimed that he was the 'father' of Bigfoot and that the 1958 Bluff Creek tracks were made by Wallace as a practical joke. A photograph of Wallace's nephew with the oversized wooden feet used to fake the footprints was carried in newspapers around the world. Meldrum has noted that even if casts that might be associated with Wallace are removed from the inventory, there will still remain enough legitimate tracks to make his case for the existence of Bigfoot. It would seem, however, that even supporters of the authenticity of the tracks realize the need for other sources of evidence.

*   *   *

3. A Bigfoot print, near Paris, Texas: 'A place for dreams,
A place for heartbreak. A place to pick up the pieces.'

As T.J. lifts the plaster cast from the mud, he turns it over and points out the
pattern of ridges on the bottom. We are not dealing with a hoax, a bear, or a
large barefoot Texan. This is a Sasquatch track.

The terrain looks nothing like the opening credits of Wim Wenders' film,
Paris, Texas, *all desolate and barren, where Harry Dean Stanton walks the*
land like a skinny Sasquatch in a business suit and red baseball cap. In the
real Paris, Texas we are standing on the shore of a swampy lagoon that
opens out into a rather large lake. Enormous lily pads dot the surface of the
water near the shore, along with tall cat-tail reeds. T.J. has described the
location as prehistoric, and that is certainly how it feels. I can easily imagine
an encounter with something primitive taking place here. I expect to see a
pterodactyl sailing overhead or the head of a long-necked dinosaur rising
from the water.

T.J. is now telling the camera how the residents of the area have long
protected the creatures that live here. They know of their existence and live
in harmony with them but seldom make any reports of their encounters
to the outside world. Driving along the roadway that approaches the lake
I had noticed evidence of their folk-art tributes to their hirsute woodland
neighbors. A plywood likeness of Bigfoot that stands in front of one home is
perhaps meant as a sign of respect and honor, like some primitive carving
of a deity meant to express devotion and to appease the god's anger. I am
reminded of the movie King Kong *where the natives of Skull Island live in*

fear of the great beast. Indeed, Biscardi has described our expedition as the 'Quest to Capture America's King Kong' and perhaps this too is what he had in mind. Of course, King Kong is only an image on film. The creature we are looking for leaves tracks in the Texas mud. That is not to say that film is not an important tool in the quest for Bigfoot, however. After all, our discovery and analysis of the footprint have taken place in front of a camera and Tom is now directing me to help place motion-activated cameras in trees all around the area. We have found his tracks, now we are going to take his picture.

4. A frame from the famous Patterson–Gimlin footage.
Is this a female Sasquatch, 'Patty,' or a pal of Patterson's in fancy dress?
The debate continues.

THE PATTERSON–GIMLIN FILM

In addition to folk tales, eyewitness accounts, and tracks, the other critical piece of evidence cited in defense of the existence of North American bipedal hominids is photography. Though many photographs and filmed images have been produced, the image usually granted the highest level of credibility is the Patterson–Gimlin footage from 1967. The well-known footage of a female Sasquatch walking along a dry creek bed while casting a glance over her right shoulder has been subjected to various types of analysis in order to determine its authenticity. As with Bigfoot prints, the options are pretty clear cut – either 'Patty,' as the creature in the film is affectionately called, is a hoax or Patty is a hairy hominid.

It was on October 20, 1967 that Roger Patterson and Bob Gimlin caught the famous image on film. Visiting northern California for the express purpose of capturing a Bigfoot on film, they both claimed to have been caught by surprise when one of the creatures walked right in front of their horses. Exploring the area around Bluff Creek that had been made famous by Bigfoot less than a decade earlier, Patterson and Gimlin rounded a large obstacle and were shocked to see a Sasquatch crouching beside the creek to their left. The creature then stood upright and proceeded to walk away from them, going from left to right. Patterson's horse was startled by the sight and he quickly found himself on the ground. He nevertheless was able to retrieve his 16mm movie camera and begin filming. Krantz describes the film in this way:

> The film begins when Patterson is at a distance of 112 feet (34 m) behind the creature, somewhat to the right of its path, and running toward it about half-again faster than it is walking. At first the film jumps badly, the small image usually being unrecognizably blurred or completely off the frame. Whenever it can be seen clearly, it is simply walking away.
>
> After Patterson had closed the distance between it and himself to just 81 feet, he stumbled to his knees but kept the camera going. About then the sasquatch turned its head (and upper body) to face him briefly while continuing to walk with long strides . . .
>
> From where he stopped, Patterson was able to hold the camera fairly steadily for most of the remaining filming. The creature's legs were obscured by stream debris in many of the best frames . . . and it was getting progressively smaller on the frames as the distance increased.

Near the end of the footage, Patterson quickly moved his position about 10 feet to the left (3 m) for a better view. The subject continued to walk almost directly away until the camera ran out of film when it was at a distance of 265 feet (80.8 m). The entire incident was over in less than two minutes, and Patterson had 952 frames of color film of the first and only sasquatch he ever saw. (89–90)

Through his analysis of the film, Krantz believed that he was able to make determinations concerning the physical characteristics of the creature. He estimated that the creature stands six feet tall and weighs around 500 pounds. He also concluded that the filmed subject walks in a manner atypical for humans. The creature exhibits the knee bends and leg extensions of a much smaller individual moving at a fast pace. It also exhibits the motions associated with a very heavy human, but with strides much longer than that normally taken by such a heavy person. He writes: 'Judging from the way it walks, there is no possibility that the film subject can be a man in a fur suit' (115).

> 6 JUDGING FROM THE WAY IT WALKS, THERE IS NO POSSIBILITY THAT THE FILM SUBJECT CAN BE A MAN IN A FUR SUIT 9

The story does not end there, however, because in 2004 an old friend of Patterson and Gimlin by the name of Bob Hieronimus claimed to have been the man in the suit. Hieronimus claims that he was promised $1,000 if he would wear a suit for Patterson's camera. The suit, according to Hieronimus, was a modified gorilla costume, with football shoulder pads to add bulk and a football helmet with a mask attached for the head. Family and friends of Hieronimus reported having seen the suit in the trunk of his car, and in some cases said they had worn the head themselves. Hieronimus claims that the unusual gait of the creature in the film was a result of the confines of the costume, including the large, fake feet that he was wearing.

GIGANTOPITHECUS AND OTHERS

Satisfying himself with the evidence from eyewitness testimony, folklore, tracks, and the Patterson–Gimlin film that there is indeed a large, hairy, bipedal hominid stalking the forests of the

Pacific Northwest, Krantz also sought evidence from the field of anthropology. Specifically, he combed the fossil record for evidence of a Bigfoot-like creature. He found a candidate in *Gigantopithecus blacki*. The evidence for this primate consists of parts of the lower jawbones from a handful of animals, and thousands of teeth. According to Krantz, *G. blacki* is believed to have lived in southern China around one million years ago. Krantz notes that he is not the first to make the connection between contemporary creatures and the giant ape. In 1968 Bigfoot expert John Green had suggested that Sasquatch might be a remnant population of *G. blacki*. Krantz, however, claimed to have gone a step farther than others by reconstructing a physical description of the prehistoric primate based on those few fossil remains. He thought that such a reconstruction showed that both the Himalayan Yeti and the North American Sasquatch might be related to the Chinese 'super-gorilla:'

> I formally proposed to equate the two species under *G. blacki*, but expressed the hope that *G. anadensis* would become the accepted name if the Sasquatch proved to be a separate species. I also suggested using the name *Giganthropus anadensis* if it should prove to be generically distinct, or *Australopithecus anadensis* if future discoveries should point in that direction. However, such proposals carry no legal weight under the established rules of zoological nomenclature. (274)

Alas, Krantz admitted that the acceptance of his suggestions is hampered by one problem in particular. Namely, 'All of these ideas presently suffer from the fact that there is no direct overlapping of evidence between the known fossils and the reported living species' (274).

Nevertheless, Krantz argued passionately for his ideas. He rejected what he called 'the common view' that little can be determined about body size and locomotion from the few simple jawbone and tooth fragments at our disposal. Indeed, he argues that upper teeth can be used to reconstruct an upper jaw to match the extant lower one. Related musculature and other structures can then be easily inferred. Finally, a solid conclusion can be reached: 'Given the great divergence of the jaw, as well as the reduced sectorial complex in the teeth, an erect posture can be deduced' (275). From there, the rest can be carefully worked out:

> The picture of an adult male *Gigantopithecus* may be summarized from these data and deductions. It was an erect, bipedal primate with the

wide shoulders and strong arms of an ex-brachiator. Its body would be ape/human-like in its broad chest, short waist, and lack of external tail. It would weigh about 350 kg (800 lbs.) and stand perhaps 2.5 m (8 ft.) tall, on legs and feet of roughly human proportions and stout design. It would be covered with normal primate hair and have a gorilla-like face. Its intelligence should be in the general area of living apes, with no cultural capacity or language. The female would be smaller, at 250 kg and 2 m, but otherwise the same.

An animal exactly fitting this description is often reported as seen in North America. (286–7)

Thus, Krantz argues that *Gigantopithecus blacki* is a viable candidate for the species identification of Bigfoot. In other words, Bigfoot, and the Yeti, may constitute a remnant population of the giant prehistoric ape.

This claim leads Krantz to consider the possibility that such creatures exist not only in the Pacific Northwest and the Himalayas, but in many other places around the globe. Indeed, reports of large, hirsute, bipedal creatures are common in many other regions of the world. Some of these creatures, such as the Almas of Mongolia, the Barmanou of Afghanistan and Pakistan, and the Nguoi Rung of Vietnam are in locations easily explicable in terms of the *Gigantopithecus blacki* hypothesis. Others, such as the Hibagon of Japan, the Kapre of the Philippines, and the Yowie of Australia are a bit more problematic. He admits that:

When it is suggested that a wild primate is found native to all continents, including Australia, then credibility drops sharply . . . Beyond a certain point, it can be argued that the more widespread a cryptozoological species is reported to be, the less likely it is that the creature exists at all. (197)

However, he does not rule out the possibility that small, remnant populations may exist around the globe, though perhaps having evolved into other closely related species or sub-species. He says that, 'Reports from states like Indiana, Ohio, and Pennsylvania seem to be increasing in number and apparent authenticity' (200), and considers the Florida skunk-ape (a smaller, smellier, more aggressive cousin to the Sasquatch of the Northwest) to be particularly well attested to by footprint and eyewitness evidence. This is also true of sightings throughout low-lying swamp areas in the lower Mississippi and other large river basins.

*   *   *

Which explains why I am standing knee-deep in the swamps of north Texas, worried about snakes and ticks and the lightning flashing overhead. A creature has been seen in this area, first by locals and then by members of Tom Biscardi's team. Tom himself claims that there is a family of creatures, what he calls a 'pod,' living on the site of the old Camp Maxey army training grounds, perhaps sheltering in underground bunkers.

I have been assigned to a group under the leadership of T.J. We are to scout out an area of woods bordering on the lake. It is to be a completely dark excursion, no lights allowed. The heavily overcast skies and intermittent rain make it impossible to see anything. T.J. directs us to sit around the base of a large tree until our vision has adjusted somewhat to the darkness. Once acclimated, we begin our trek. We walk in single file, T.J. leading the way.

Not only is our expedition dark, it is also silent. T.J. directs our movements with hand signals, barely visible in the darkness. Every few minutes he signals for us to stop. After we all drop to one knee, I do as I have been instructed and scan 360 degrees with the night-vision goggles that I have been assigned. A wave of my hand indicates that all is clear, a thumbs-up is reserved for a sighting. As soon as I give my signal we are up and moving again. Now, however, I have a problem. The night-vision goggles have ruined my own natural night vision – I can't see a damn thing. We are leaping over a deep muddy trench, or at least the others are leaping; I am struggling through the knee-deep water, trying to keep the expensive goggles dry.

Soon the wind begins to pick up and the rain grows more intense. Lightning flashes overhead. The weather forecast predicted possible tornados, with dangerous lightning and hail. I am grateful for the poncho and big Texas cowboy hat that I am wearing, otherwise I might drown standing up. If the creature is out in this weather, he is a damn fool – like me. This is what Bigfooting is about, however. Eyewitness accounts, footprints, and videotape just aren't enough; we need a specimen. Which is why we are here tonight, in foul weather and pitch-black darkness. Our team is swinging to the right, another team is swinging to the left. If the creature is here we will drive it right into Tom Biscardi's lap.

FROZEN MAN

Despite the evidence mounted by Krantz, and later by Meldrum, the existence of Bigfoot still remains uncertain. The evidence from folklore, eyewitness accounts, footprint analysis, film, and the fossil record fails to convince a vast number of people. Clearly what is needed is a specimen, living or dead (as long as the corpse is less than half a million years old). The flesh and bones of the creatures decay quickly in the woods, we are told. It is not surprising that no one has stumbled upon the remains of a recently deceased Sasquatch. But wouldn't you think that someone would have found something by now? Perhaps someone did.

According to Loren Coleman's account, the body of one of the creatures may have been encased in a block of ice in the fall of 1967. That was when Terry Cullen paid a quarter to see such a sight on exhibit in Milwaukee. It was on display in a refrigerated glass coffin shown by exhibitor Frank Hansen. Cullen notified the Bigfoot community and soon Bernard Heuvelmans and Ivan Sanderson were examining the creature. What they found was the likeness of an adult male with large feet and hands. It was covered with dark brown fur. Both eyeballs were missing from the sockets, one entirely absent and one dangling on the face. It also had an open wound and broken bone in its left arm. The smell of rotting flesh was obvious.

Heuvelmans described his examination:

> The specimen at first looks like a man, or, if you prefer, an adult human being of the male sex, of rather normal height (six feet) and proportions but excessively hairy. It is entirely covered with very dark brown hair three to four inches long. Its skin appears waxlike, similar in color to the cadavers of white men not tanned by the sun ... The specimen is lying on its back ... the left arm is twisted behind the head with the palm of the hand upward. The arm makes a strange curve, as if it were that of a sawdust doll, but this curvature is due to an open fracture midway between the wrist and the elbow where one can distinguish the broken ulna in a gaping wound. The right arm is twisted and held tightly against the flank, with the hand spread palm down over the right side of the abdomen. Between the right finger and the medius the penis is visible, lying obliquely on the groin. The testicles are vaguely distinguishable at the juncture of the thighs. (Heuvelmans in 'Bulletin of the Royal Institute of Natural Sciences of Belgium,' quoted in Coleman 2003b, 113)

Unfortunately, before the researchers could progress any further the creature disappeared, pulled from public viewing by Hansen. When Hansen did resume the exhibition of the creature, it was reported that he had replaced the original with a wax likeness. In other words, Hansen's exhibit turned out to be a gaffe – a fake designed for the sideshow. Heuvelmans and Sanderson, however, continued to claim that the specimen they had examined – the original Frozen Man – had been the real thing.

TO KILL OR NOT TO KILL

With the passing of time and the continued skepticism of the scientific community regarding the evidence presented by Krantz and others, many Bigfoot believers have come to the conclusion that the only way to prove the existence of the creature is to kill or capture a specimen. Even photographic evidence, in this digital age, can be easily manipulated. Waiting for another Frozen Man to show up at a sideshow is likewise a hopeless proposition. A specimen is required – not footprints, not eyewitness testimony, not references in folklore, not photographs or videotape, not prehistoric fossils of similar creatures – but a body, either living or dead. Indeed, some have said that, considering the problems inherent in capturing such an unpredictable creature, the researcher must be prepared to kill in lieu of live capture if push comes to shove. If we are dealing with a remnant population, a creature that is obviously endangered, then perhaps one of its number must be killed in order to save the rest.

Such a pragmatic attitude has been criticized by Russian Bigfooter Dmitri Bayanov, and the debate has proven to be a heated one in Bigfoot circles. Bayanov has proposed a method of baiting and habituating the creatures in a way that would allow the researcher to capture solid proof of their existence on film. Others consider this approach naïve. For example, long-time Bigfoot researcher John Green has argued that the collection and dissection of a Sasquatch specimen is crucial to the scientific process. Speaking before the 2003 International Bigfoot Symposium, Green noted that once professional scientists get involved,

> They will want to study these creatures in every possible way, and some, I expect, will get official permission to collect for dissection not one, but several.
> I am, of course, familiar with the argument that we should only study sasquatch the way Dr. Goodall studies chimpanzees, and such methods will certainly be tried. I doubt, however that they will prove to be practical with creatures that are so much more mobile in their home environment than humans, and even if they are practical, they cannot provide all the information that will be wanted.

The anatomy of creatures that walk in much the same way that humans do is going to be studied in detail. To do that effectively will involve dissection, and will require more than one cadaver, because the cuts made while exploring one bodily system destroy the others . . . (Green, as quoted by Craig Woolheater at www. cryptomundo.com)

Green's comments have met with strong disavowals from others in the Bigfoot community. Loren Coleman, for example, responded to Green with the following comments:

The first large unknown hairy hominoid captured will live its life in captivity, no doubt, and there it may be examined internally. MRIs, CAT scans, EKGs, and a whole battery of medical and other procedures may be used to examine it.

It is doubtful the first one will be returned to the wild, so, of course, it will die someday within the reach of future scientific examinations. Then it will be dissected, just as newly discovered animals, including various kinds of humans, have been for further study. But in the meantime, why not study the living animal's captive and adaptive behaviors?

The days of Queen Victoria, when only killing an animal would establish it was real and not folklore, are, indeed, long gone. (www. cryptomundo.com)

Bayanov's reaction to Green and others is equally forceful. He argues that the Sasquatch should not be considered as merely an animal, but as something altogether closer to human and thus worthy of special consideration:

I can imagine young Sasquatches doing very well in a school especially designed for them. If a chimp brought up and taught by humans can acquire a certain vocabulary, I wonder what heights of scholarship can be attained by an aspiring young Sasquatch under human tutorship. If a human child brought up by animals becomes an animal, I wonder what will become of a homi child brought up by humans.

That's one side of the matter. The other is a reversal of roles, with the Sasquatch becoming a teacher of human boy and girl scouts in the art of survival in the wild. With the present day 'Back to Nature' trend I regard such a possibility as quite feasible. In that case we are bound to have parent–teacher conferences with a somewhat different agenda and composition. (www.bigfootencounters.com)

The debate does not seem likely to end anytime soon.

*   *   *

I am Tarzan, stalking the bolgani through heart-of-the-jungle darkness.

My team, under T.J.'s leadership, is hiking through snake-infested waters during a terrible thunderstorm in pitch-black conditions. We are, theoretically, driving the creature into a trap. If all goes according to plan, Tom Biscardi will be able to capture the creature and put an end to the mystery of Bigfoot once and for all.

Though this is an expedition intent on capturing the creature, Biscardi does not rule out killing it if we have to. A regiment of Bigfoot hunters from Arkansas has joined Tom's group for this hunt, all of them carrying rifles. In addition, Tom has placed one of our team in a sniper position. Despite the wind and lightning, a member of the team has climbed to the top of a tall pine tree. He sits there, crouching on a hunter's platform with his rifle at the ready. Assuming he can see well enough in the darkness, he has orders to shoot the creature should it place any one of us in danger, or should it be on the verge of escape. As we move closer to the sniper, T.J. breaks his own rule of silence and radioes our position. He doesn't want one of us to be mistaken for the creature and fall victim to the sniper's bullet. The radio is not working properly, however – he can't raise the gunman in the tree. As an alternative, he radioes our position to the command post and asks them to contact the sniper for us. Without waiting for an answer, we march ahead. The hell with lightning, snakes, and huge hairy beasts – I'm most scared of the guy with the gun, fearing he will mistake us for the resident family of Bigfoot, on our way home to the bunker after a night of foraging and terrorizing the locals.

Fortunately, we all make it back to the command post safely. There have been no sightings of the creature. The weather is just too bad. The Searching for Bigfoot team huddles around our trucks. I pass a flask of bourbon around and watch as one of our guys uses it to wash down a handful of pills.

I decide to call it a night, or a morning since it is only three or four hours to sunrise. I drive the long road back to the hotel, change out of my wet clothes, and crawl into bed. Just as I do, the phone rings. It is Tom Biscardi.

'Greg, you've got to do me a favor. Go next door and wake up the cameraman. Tell him to get back out here as fast as he can. The guys from Arkansas just radioed in. They've got a bitch up a tree and she's throwing rocks. We're going to them now and we need a camera.'

'Sure Tom, I'll get him.'

Apparently the cameraman had followed my lead and called it an early night; unlike me, he was not answering his phone. He does answer the door, however.

'Tom needs you and your camera back out in the field,' I tell him. 'The Arkansas boys have radioed that they have a bitch up a tree and she's throwing rocks.'

'What the hell does he want me to do about it?' he asks, 'Stick a camera up her ass? Why don't they just shoot the bitch?'

He closes the door and, I assume, goes back to bed.

I go to my room and do the same.

The next morning, the news isn't good. Apparently the boys from Arkansas ran out of bullets before the bitch ran out of rocks. (How did she get rocks up the tree? Are Bigfoot marsupials? Did she carry them up in her pouch?) Other members of her family arrived and began threatening the hunters. They were forced to retreat – without a specimen, a footprint, or a video.

❝ THE HELL WITH LIGHTNING, SNAKES, AND HUGE HAIRY BEASTS – I'M MOST SCARED OF THE GUY WITH THE GUN ❞

*   *   *

I saw strange footprints in the north Texas woods. They were as clear as day in the moist soil. They were not bear tracks and they were much too large to be human. They were, I have no doubt, Bigfoot tracks.

Bigfoot tracks are one of the central elements in the story of North America's giant hairy hominid. Since that day in 1958 when Jerry Crew made headlines with his plaster cast of a giant footprint, the case for the existence of Bigfoot has centered around these big prints, found from Washington State to the southern tip of Florida. If these tracks are real, the argument goes, then Bigfoot is real.

Tracks are not the only bit of evidence that has grown up to support the existence of the creatures, of course. There are other things. Evidence is found in folklore, old newspaper stories, and eyewitness accounts. People of sterling background and with excellent reputations swear that they have seen these forest giants. There are photographs of the creatures, and moving pictures. Footprints themselves are subjected to detailed analysis, revealing dermal ridges and pressure points. Some people claim to have examined a carcass encased in ice. Others actively hunt the creature – some to capture, some to kill – all looking for evidence, for proof, that the creature is real. The fossil record has been carefully studied for signs that something like Bigfoot once walked the earth, for if he was once real he might be real now as well.

The hunt for Bigfoot has taken on the trappings of science. Once the footprints were encased in plaster, available for study and analysis, there was no going back. Bigfoot moved from the world of folklore and tall tales to the world of evidence and proof. The tracks are a metaphor for what happened to the stories themselves. It wasn't just the tracks that were captured in plaster and subjected to the microscope, but the legends. Hairy giants became Sasquatch; Sasquatch became Bigfoot; Bigfoot became *Giganthopichecus*.

But it all goes back to the tracks.

Like the tracks I saw in Texas, the Bigfoot tracks of Paris.

Either the tracks that I saw were really the tracks of some unknown creature with very large, human-like feet, or they were fakes. My rational mind reminds me that the tracks seemed too perfect, that they were found in a location that was far too accessible and open. Dermal ridges or not, they looked like fakes to me.

If they were fakes, then of course we wonder who was responsible for them and what their motives might have been.

It is possible, of course, that Biscardi and company were fooled by a third party, maybe a local resident who thinks it is all very funny. But the tracks looked too good to be just a practical joke. They looked fake to me, but they looked like professional fakes. I suppose that someone else in the Bigfoot community could be responsible. But why would anyone bother? Biscardi claims to find tracks like this all over the country. Surely he is not being followed around by a merry prankster, hiding in the bushes and snickering into his sleeves. (Hmm, Sasquatch is sometimes portrayed as a trickster figure in Native American lore. Perhaps it is a Sasquatch that is following Biscardi with a phony wooden foot, leaving faked footprints in obvious places and then laughing about it all the way back to the bunker.)

The logical explanation is, of course, that Biscardi is somehow involved in the hoax. The question is 'Why?' Biscardi does have a reputation for fakery in some quarters of the Bigfoot community. He is associated with Ivan and Peggy Marx, for example, regarded by some as the producers of some ridiculously camp photographs and video of the creature. He also once claimed on Coast to Coast AM, the paranormal talk-radio show, that he had a creature in captivity. A logical motive might be money, but I have some idea that Biscardi has shelled out a good deal of his own money on this expedition. Some of the guides told me that he paid for them to fly in for the event. The fee he charged me is minimal and it included my hotel room and at least one meal. All things considered, I don't think it possible that he made a lot of money on this, or that he ever makes a lot of money from Bigfoot. It is of course possible that Tom Biscardi and his crew faked the prints for reasons other than money, perhaps simply to impress me and the local news media or to perpetrate some sort of elaborate joke. But this only raises the question of why anyone in his position would want to do this. I know that there are easier ways to get on the local news. I also know that there are easier ways to impress the likes of me. And after fifty years, the joke just isn't that funny anymore.

This is funny, though – funny strange, that is.

When I knelt in the mud and examined the footprints I felt a tingle of excitement. I did not stop to ask myself who faked the prints. I did not even stop to remind myself that they were clearly fakes. These thoughts came later, in moments of reflection.

In that moment when my hand touched the sandy soil, when I gazed out across that primeval lake, when I thought about the night's hunt that was soon to follow, in that moment they were real. It made no difference how they came to be there. They *were* Bigfoot tracks.

Two

Cryptozoology

A game warden is carefully and forcefully explaining the dangerous conditions we will face if we choose to continue with our plans and spend the night exploring the forest across the lake in an attempt to capture Bigfoot. With every word she speaks I am questioning the wisdom of crossing the lake with this group of people – people that I have only just met, mostly visitors like myself to North Texas, many of whom, unlike me, appear to be heavily armed.

First, she tells us about the snakes. There are water moccasins, which thrive in and around the lakeshore, as well as copperheads and rattlesnakes. A bite from any of these could be painful and potentially deadly. The woods are also the habitat for wild hogs. As large as 400 pounds, these animals are especially dangerous at this time of year because many of them have new litters. Under no circumstances is it a good idea to find oneself in the position of being between a female and her young. In addition to snakes and wild hogs she also cautions us about unexploded munitions. The area had, at one time, been used as a military testing range and is littered with unexploded shells. The warden instructs us not to touch anything that looks suspicious, but instead to find some way to mark the location of the suspected ammunition and report it to the authorities for removal or safe detonation.

Next, the warden gives our group an updated weather forecast. The night will bring strong storms with heavy rain, high winds, and possible hail and tornados. She instructs us to make certain that the boat is sheltered before the storm hits, otherwise it is sure to be swamped by morning. Finally, the warden reminds us that we are an hour from the nearest medical help back in Paris. If we call for help it will be at least two hours before we can be anywhere near a hospital. That, of course, is assuming that our cell phones will work on the opposite side of the lake and that we will be able to cross the lake to reach the ambulance.

That does it for me. Snakes, wild hogs, military explosives, severe weather, and armed strangers are conditions that might be surmountable if there was the possibility of rescue, the possibility of paramedics or Texas Rangers arriving to set things right, just like in those old 1970s TV shows. In an isolated location, however, things could go pretty badly. The smart thing to do would be to stay at base camp and monitor the images from the video camera placed on the other shore earlier in the day. I should be safe as long as I stay at the command post. I call my wife and tell her not to worry. With distant lightning flashes causing the cell phone to crackle, I let her know that I am safe, that I won't get into a boat, and that if the weather gets bad I will head back to town.

Then the shouting starts. Something just moved in front of the camera. Whatever it was, it looked big, certainly bigger than any hog.

'Get a team over there! Now!' Tom Biscardi shouts.

I hear the boat's motor start and feel myself grabbing my backpack and running to join the group that is climbing aboard the small fishing boat. I have to wade out several feet to get to the boat and my boots, purchased for this trip and promising to be waterproof, quickly fill with water. Even here, right by shore, it is already pitch dark. Those of us with flashlights point them toward the front as the boat begins to cross the lake, providing what little extra light we can. Soon we are at the other shore and piling out of the boat. Again, the water rushes into my boots. I clutch my backpack, pull the hood of my poncho tighter and wade ashore through the high reeds and driftwood, thinking now only of the water moccasins, flashing my light in the water, looking for any sign of their cotton-colored mouths. Just as I reach dry ground with the other people in the party I hear the boat begin to pull away. It is going back across to the other side, leaving us to face the snakes, pigs, artillery, and the weather. I only hope that they remember to get the boat out of the water before the storm comes. It will be no good to anybody if it swamps.

As we reach dry ground, I begin to ascertain who my companions are. We all rushed into the boat so quickly, our bodies and heads covered with ponchos, that I could not at first tell who I was with. This seems to be a concern for everyone, because as soon as we find a little shelter from the rain under the forest canopy we stop and huddle together — introducing ourselves and planning our mission. The leaders of our group are experienced Bigfoot trackers named Hondo and Hawk. They will be in charge of the guns. The other four of us are Bigfoot amateurs. Snake Man is a Paris local who has been invited along because of his work as an animal wrangler. When *folks have trouble with wild animals, especially snakes, they call him for capture and release. He is equipped with a snake-bite kit and a fair amount of experience in the woods of north Texas. Andy is from New Jersey, a large man without any real outdoor experience. He is a lifelong Bigfoot believer, who says that he has been dreaming of this moment for most of his adult life. He is inexperienced, but seems ready for anything. Tiffany and Jennifer are psychic researchers, along on the expedition to assist in the use of special equipment: night-vision goggles and extra-low-frequency sound receivers designed to pick up the low rumblings used by Bigfoot to communicate with one another. A perfect cast for a low-budget horror movie. Unfortunately, in movies like this, my character is usually the first to get taken out.*

We stand in the rainy darkness, waiting for our eyes to adjust to the near total absence of light, waiting for any living things in the woods to become acclimated to our presence, then we begin to move through the woods, looking for any signs of our prey – large mounds of human-like scat, small trees bent so that they crisscross and form the letter 'X', and, of course, footprints. We listen for rustling, for grunts, for low and barely audible sounds, and for the sounds of tree-knocking – a form of communication among the creatures we are hunting.

In addition to the wind, rain, lightning, and thunder, tracking our quarry is made difficult because of the shrewdness of our prey.

'These creatures are excellent mimics,' Hawk informs us. 'They can sound like almost anything: cats, dogs, owls, you name it. Any sound you hear tonight, no matter how normal it may seem, could be a S'quatch.'

I have never heard the term 'S'quatch' before, though I assume it is a contraction of the term 'Sasquatch,' but I do understand the implications of Hawk's remark: our search for Bigfoot is not going to be easy. They could be anywhere.

As if to reinforce this point, a radio message comes in from the command post: the video camera on our side of the lake has just stopped transmitting. Can we check it out? As we begin to walk slowly toward where we hope the camera is located, a second message comes through.

❝ THEY CAN SOUND LIKE ALMOST ANYTHING: CATS, DOGS, OWLS . . . ANY SOUND YOU HEAR TONIGHT . . . COULD BE A S'QUATCH ❞

'The creature is on the move! The creature is on the move!' the radio blares.

Another scouting party is located down the lake from us at a site accessible by land. They have reported that they have been charged by a large bull

S'quatch that is now headed in our direction. We can do nothing but trudge on to check the camera, which we find intact but inoperable. The rain begins to fall in sheets. Lightning flashes. The wind howls. Then another series of messages: the other team has just spotted a tornado, also headed in our direction; the water is too rough for the boat to come and get us; command post is seeking shelter – tents are blowing away – we should find a sheltered spot and hunker down; they will contact us and get us back across the lake just as soon as the weather breaks. But, of course, we don't want to get back when the weather breaks – when the weather breaks we will be fine – we want to get back now!

We find a spot somewhat sheltered from the wind, settle in, and pull our ponchos tight against the rain. Hawk, Hondo, and Snake, amazingly, are able to start a fire despite the steady rain. We all gather around, sharing our food. After what seems like an eternity the rain begins to subside. Our clothes actually begin to dry. In the distance we hear the sound of coyote, or of S'quatch impersonating coyote. Up close there is rustling among the trees. It is now well after midnight and I feel suddenly exhausted, the adrenaline going out of my system and leaving me drained. We begin to talk about where we are from and why we are here, about our own experiences, or lack thereof, with wilderness and Bigfoot and with other things.

'You know,' Hawk tells us, 'Bigfoot is not the only strange thing in these woods. These woods are full of Wendigo.'

I know that word. Some describe it as a Bigfoot-like creature with long pointed ears and dog-like snout, more like a werewolf than a big ape. Some claim that they possess supernatural powers and say that they are shape-shifters, riders on the storm, arriving with a high wind and a clap of thunder, to take their victims screaming into the night. Others say they are simply flesh-and-blood animals, huge bipedal primates, cousins to Bigfoot – with snouts like baboons and slobbering mouths filled with teeth. There is an old horror story about these creatures that I remember reading many years ago. A small hunting party, separated from their camp and the rest of their group by a lake, had encountered the legendary creature . . .

'The storm is over,' someone observes, 'the boat should be able to cross the lake now.'

But, of course, when the radio call is made, the news is bad – the boat has been swamped and they won't be able to get across until morning.

WENDIGO, SKINWALKERS, AND MAN-WOLVES

In Algernon Blackwood's story *The Wendigo*, originally published in 1910, a pair of hunters encounter a creature in the Canadian woods that they had hoped was only a legend. Blackwood's Wendigo is a mystical creature, representing the lure of the wilderness. Its voice resembles the sounds of the wild – falling water, wind, the call of birds and beasts. Grasping its victim, the Wendigo runs at such speed that the poor victim's eyes bleed and his feet burn. It can take great leaps or run along the tops of trees. It will fly, repeatedly dropping its prey, as a hawk drops a fish, over and over again until it is dead. If the victim of the Wendigo does return to civilization, it is as a shattered shell of the individual, his mind crushed by the terror of the dark woods.

As with many of the creatures brought to life in the pages of gothic horror fiction, Blackwood's Wendigo is inspired by a beast already present in folk mythology. In this case, Blackwood draws upon the folk tales of the Algonquian tribe of Native Americans. In some versions of these tales, the Wendigo is an evil spirit that influences weak and susceptible individuals to engage in horrific behavior. Specifically, when facing extreme conditions and starvation, the Algonquians believed that the spirit of the Wendigo may transform an individual into a cannibal. The victim then becomes a Wendigo. Steve Pitt describes aspects of the folkloric Wendigo in *Legion Magazine* and describes the creature this way:

> It is something like a werewolf on steroids. It stands more than six metres tall in its bare feet, looks like a walking corpse and smells like rotting meat. It has long, stringy hair and a heart of ice. Sometimes a Windigo breathes fire. It can talk, but mostly it hisses and howls. Windigos can fly on the winds of a blizzard or walk across water without sinking. They are stronger than a grizzly bear and run faster than any human being, which is bad news because human flesh happens to be a Windigo's favourite food. A Windigo's appetite is insatiable. Indeed, the more it eats, the hungrier it gets. (January/ February 2003)

It requires a leap of imagination, akin to the leaps of the Wendigo itself, however, to go from the cobwebbed pages of a gothic horror story and the campfire tales of tribal folklore to a real-life flesh-and-blood creature haunting the woods of north Texas. This claim, however, is not made by Hawk alone. For example, Loren Coleman suggests that Native American folklore has a layer of truth that should not be dismissed. Stories of Wendigo, he claims, may contain information about actual encounters between native peoples and a group of large, hairy hominids. He writes:

> In eastern North America, a specific variety of manlike hairy hominid allegedly exists which exhibit aggressive behavior, hair covering the face in a mask like fashion, occasionally piebald coloring, an infrequent protruding stomach, and distinctive curved, five-toed splayed footprints. They are inhabitants of the northern forests. (Coleman, *Crypto*, 'Hominology Special Number 1,' April 7, 2001)

Though these creatures were called by many different names by the various tribes, they are, according to Coleman, the beings described by the Algonquians as Wendigo. In Coleman's examination of native folk tales a picture of the Wendigo begins to develop that looks more like a flesh-and-blood primate than like the mystical otherworldly being of Blackwood's tale. The Wendigo that Coleman describes are large, hairy creatures with big hands and bear-like faces. They wear no clothes and do not employ the bow, instead using stones for weapons – hence the origin of the term 'Stone Giants.' They are cannibals, though it is unclear whether this means that they eat members of their own group or that they eat human prey.

Coleman suggests that accounts of human encounters with Wendigo can be found not only in native folk tales but also in stories told by early European settlers to North America. One such European encounter even comes from the pre-Columbian period:

> Leif Erickson wrote in 986 A.D. of encountering monsters that were ugly, hairy, and swarthy with great black eyes. In 1603, on Samuel de Champlain's first voyage to eastern Canada, the local natives informed the explorer of the *Gougou*, a giant hairy beast that lived in the northern forests and was much feared by the Micmac. A *London Times* article of 1784 records the capture of hair-covered hominids by local natives, near Lake of the Woods, in south-central Canada. In the *Boston Gazette* of July 1793, a dispatch appeared from Charleston, South Carolina, May 17, 1793, concerning a creature seen in North Carolina. The account centers on Bald Mountain, where the local residents call it *Yahoo*, while the Indians give it the name of *Chickly Cudly*. (Coleman, *Crypto*, April 7, 2001)

Coleman, then, not only brings the Wendigo out of fiction and folklore to stalk the forests of eastern Canada, he also sees the creature as a biological creature, an undiscovered primate. He also brings the creature closer to Texas.

The indigenous peoples of Canada were not the only Native American groups to tell stories of hairy, human-like, sometimes supernatural, creatures. Navajo folklore from the American Southwest, for example, tells of supernatural creatures known as Skinwalkers. Skinwalkers are shape-shifting witches able to assume the attributes of various animals, especially canines. They walk in the skins of animals. Just as with the Wendigo, Skinwalkers also make appearances in contemporary accounts, most notably in *Hunt for the Skinwalker* by Colm A. Kelleher and George Knapp (2005). These authors relay several interesting accounts of modern-day Skinwalker encounters, including the following:

> One story told on the Navajo reservation in Arizona concerns a woman who delivered newspapers in the early morning hours. She claims that, during her rounds, she heard a scratching on the passenger door of the vehicle. Her baby was in the car seat next to her. The door flung open, and she saw the horrifying form of a creature she described as half man, half beast, with glowing red eyes and a gnarly arm that was reaching for her child. She fought it off, managed to pull the door closed, then pounded the gas pedal and sped off. To her horror, she says, the creature ran along with the car and continued to try to open the door. It stayed with her until she screeched up to an all-night convenience store. She ran inside, screaming and hysterical, but when the store employee dashed outside, the being had vanished. (41)

According to one contemporary Navajo interviewed by Kelleher and Knapp,

> A skinwalker is the size of human, six foot and under. They don't come in most of the time to where the animals are at. They come in to where people are at. They come right here and you'd never know he was standing here looking at you in the middle of the night . . . They can take the shape of anything they want to take the shape of. (48)

The shape-shifting ability of the Skinwalkers can result in quite dramatic incarnations. For example, Kelleher and Knapp report an encounter with two Skinwalkers they describe as 'so unusual, so outside our concept of reality as to be almost comical, like something out of a Saturday morning cartoon' (49). In this case the Skinwalkers were in the form of humans with dog heads. They were also smoking cigarettes. Kelleher and Knapp's Skinwalker, it seems, crosses back into the realm of the supernatural, back into the worlds of gothic horror and folklore, but is nevertheless treated as fact, not fiction.

This bizarre shape-shifting tendency has also been associated with 'The Beast of Bray Road,' a contemporary werewolf-like creature from Wisconsin documented by Linda S. Godfrey. In one case an eyewitness describes watching such a transformation take place:

> It appeared then to be an overly large dog, or something like that. And then, its legs started moving real fast. The closest thing I can think of to describe it is when you see a person break-dancing, when they're spinning and kicking. And I was standing there trying to make sense of that, and I wondered if it was two dogs, uh, fornicating. And then, the only way I can describe it is that it was . . . morphing, and when it stopped, it turned and looked at me, and it had this dark, hairy body but the head and face of a gorilla. (2006, 4–5)

Godfrey admits to being a bit confused by such reports of supernatural behavior on the part of the beast, especially when those reports are compared with the more typical reports, which usually give the beast very flesh-and-blood characteristics. As Godfrey describes one of the earliest encounters with the beast, its need for food, in this case roadkill, is quite apparent. She describes the experience of her witness in this way:

> As she drove forward and saw the creature from the front, its position was what got her attention, however. 'It was kneeling!' she said. 'Its elbows were up, and its claws were facing so that I knew it had claws,' she added. 'I remember the long claws.'
>
> The claws had a chunk of what looked like a dead and flattened small animal: roadkill. She had the impression the creature had been dining upon it. Its eyes reflected the glow of her headlights, as most animal eyes will at night, and she was surprised when the creature didn't run away but turned its head to gaze back at her. She described the animal as 'dark brownish-gray' and the size of an average man, maybe five foot seven and 150 pounds. (2003, 7)

The creature also seems to enjoy scavenging from garbage cans. Godfrey recounts one eyewitness report of two large animals prowling through garbage cans under the glare of a streetlight. The creatures had ears like a German Shepherd, long sinewy bodies, and oddly small feet. They were covered in pale grey fur and had long snouts with black noses (94).

Despite these reports of animal-like behavior – behavior that seems to indicate a natural explanation for the beasts – Godfrey's files are filled with stories, like the story of the shape-shifting break-dancer, which seem to indicate a more supernatural explanation. Some witnesses insist that what they have seen is an odd, but natural, creature. Others describe it in demonic and supernatural terms. Sometimes, Godfrey notes, what witnesses report cannot easily be classified into either category, but has characteristics of both natural and supernatural phenomena. 'Sometimes,' she writes, 'it is difficult to separate the two categories, as the border between them occasionally turns porous, allowing things to seep fluidly between them' (xii).

CHUPACABRAS

The blending of natural and supernatural elements found in accounts of Wendigo, Skinwalkers, and the Beast of Bray Road is also a prominent characteristic of stories concerning the Latin American creature known as Chupacabras, the 'Goat Sucker.' Starting in the 1990s in Puerto Rico, stories of the Chupacabras quickly spread to other nations, including the United States. The Goat Sucker acquired its name because of its tendency to leave farm animals drained of blood, with only a small puncture wound providing any evidence of how this might have been done. The killing of farm animals often corresponded with sightings of a strange creature – usually described as between three and four feet tall, with powerfully built hind legs (like a kangaroo), and dark black eyes that glowed red at night. The creature was also described as having a strange crest of spiny feathers running down its back. In addition to these strange, but perfectly natural, characteristics, witnesses also claim to have experienced things that would seem to indicate that something more than a flesh-and-blood animal was involved – including some witnesses who claim that the creature emitted beams of light from its eyes that brightened the surrounding darkness like flashlights.

Scott Corrales, in his book *Chupacabras and Other Mysteries*, provides the most complete record of the early reports from Puerto Rico and it is clear that those reports contain a bizarre mix of both natural and supernatural elements. For example, Corrales describes witnesses' accounts of the creature in flight. Some accounts indicate an unusual but natural phenomenon, while others indicate something weirder:

Many who saw the Chupacabras said it has a web of skin connecting its wrist to its knee or ankle, that this web forms a 'wing,' like that of a flying squirrel when it raises its arms, and that this structure allows it to glide like a hang glider. But some witnesses insisted that the Chupacabras has a levitation capability that allows it to float through the air like Superman, in level flight without flapping wings, and without other means of propulsion. One witness claimed that the extremely rapid movement of small, feather-like appendages along its backbone propelled it like a bumblebee. (4)

Whatever it was, natural or supernatural, the Chupacabras quickly became the topic of conversation throughout Puerto Rico. Soon, people everywhere were seeing the Goat Sucker. Corrales reports that in May 1995 a San Juan television station reported that a police officer and several witnesses had seen a gargoyle-like creature while waiting at a bus stop. The creature was reportedly spotted outside a government building unpleasantly devouring a large rat. The police officer attacked it with his baton, but quickly had the tables turned on him. The Chupacabras flew into the air and grabbed the baton in its claws before flying away. Soon after, the creature was spotted flying over a busy expressway (15).

> ❝ SOME WITNESSES INSISTED THAT THE CHUPACABRAS HAS A LEVITATION CAPABILITY THAT ALLOWS IT TO FLOAT THROUGH THE AIR ❞

The strangeness was only getting started.

One witness spotted the creature outside her home and managed to get a close look at it through a window. She reported:

It was such a weird creature that I even got down on the floor to see if it had genitals. It had nothing at all – it was 'plain' and sealed. I laughed, and said to my mother, 'What the heck is this? Does it defecate through its mouth after it eats?' It made robot-like movements as if being controlled by someone. (35)

Corrales also recounts this encounter between another police officer and the creature:

The creature attacked a Chow dog at policeman Juan Collazo's home. He heard some noises in the lower part of his home. He went down with his service revolver, and he saw the creature attacking his dog.

He immediately fired at it. His car was parked behind the creature. The creature took the bullet's initial impact, bounced off the wall, took off in a flash, and disappeared. He says it flew . . . The creature had apparently shaved the place on the dog where it was going to make its incision. (57)

Corrales offers up several possible explanations for the origins and nature of the creature, including that the creature may be an extraterrestrial, a visitor from another dimension, or the result of some bizarre genetic experiment. In any event, whatever it may be, Corrales argues that it must be taken seriously. '*Real* animals belonging to *real* people *are* being slaughtered by a being not native to the Puerto Rican ecosystem' (164).

It did not take long for the Chupacabras to move beyond the Puerto Rican ecosystem, however. Fairly quickly after the initial sightings in 1995 the Chupacabras appears to have migrated to other regions of Latin America and to the United States. Reports of the creature surfaced in southern California and Miami, Florida. Within a few years the creature was making news and headlines in Texas. Everywhere it appears, reports of dead farm animals occur, their blood drained from their bodies. As with Bigfoot, the only hope for

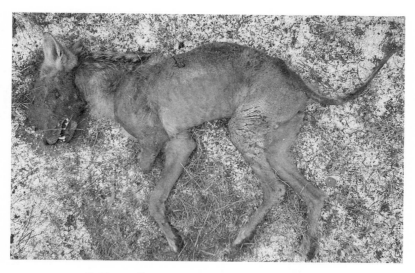

5. Texas Chupacabras or coyote with mange?

solving the mystery was to obtain an actual physical specimen. At this point, Chupacabras hunters caught a break.

Texas television stations reported that residents had discovered the remains of the mysterious creature – not once, but several times. Video and photographic images of the carcasses of strange creatures were broadcast on television and soon spread on the internet. These creatures were roughly canine in appearance, but completely hairless and with powerful back legs. Their teeth were extremely long, growing from protruding gums. Farmers reported that in the preceding months chickens and goats had been found drained of all blood. Were these the mysterious Chupacabras? Veterinarians and experts claimed that they were coyotes suffering from severe cases of mange and malnutrition. Skeptics scoffed at the claims, as did true believers in Chupacabras, who noted that these animals, strange though they might be, were certainly not the same creatures that had been seen by other witnesses. The original creatures – creatures with high-beam eyes and the ability to fly – could certainly not be mistaken for mangy coyotes. The mystery, even with a specimen, was far from over.

MOTHMAN

Chupacabras was not the first mysterious creature to thrive both on land and in the sky. In 1966 Point Pleasant, West Virginia became the center of a series of sightings immortalized by John Keel in his best-selling account *The Mothman Prophecies* (later made into a movie). The sightings began in November with the report of an 18-year-old girl, Connie Carpenter, who reported that, while driving past a deserted golf course at around 10:30 a.m., she observed a huge gray creature, shaped liked a man but considerably larger. Like the Chupacabras, the eyes of the creature made quite an impression: they were large, round, and glowing red. Keel reports that as Carpenter slowed down for a closer look, the creature unfolded a pair of huge wings. It then rose straight up into the air, again like the Chupacabras, more like a helicopter than a bird. It swooped low over her car before she sped away in terror. Keel reports that during the winter over 100 people saw the creature, dubbed the Mothman.

A group of teenagers described the entity as human shaped and six or seven feet tall. It had huge wings folded across its back. Its enormous eyes glowed like automobile reflectors. It was gray and walked on two legs. As they sped past in their automobile, Mothman

6. The Mothman Statue, Point Pleasant, West Virginia, by local sculptor Robert Roach. Not something you'd want to see when the mist rolls in.

spread its 'bat-like' wings and rose straight up into the air. According to the witnesses, the creature followed their car, traveling at 100 miles an hour, without ever flapping its wings. It squeaked like a large mouse. They only escaped because the creature turned away as they drew near to town. They could not help but notice that a large dog carcass that had previously been seen on the side of the road was now gone, perhaps taken by the creature, who, like the Chupacabras, became known for its taste for small animals. Keel describes one of his victims, a dog: 'There was a very large, very neat hole in its side, and the animal's heart was lying outside the body. It looked as if something chewed it out. There were no other marks on the body' (266). Other than his victims, the Mothman never left behind any other physical evidence. The terror from the sky doesn't even leave footprints.

BIG BIRD AND ROPENS

Ken Gerhard, in his book *Big Bird! Modern Sightings of Flying Monsters*, examines eyewitness accounts of giant bird sightings, the best known being the report of 9-year-old Marlon Lowe of Lawndale, Illinois. According to reports, in 1977 Marlon was snatched from the ground by a giant black bird, in sight of his mother and other witnesses. He was carried several feet before his struggles freed him and allowed him to drop to the ground (61). Marlon maintains to this day that the event was neither a dream nor the product of his overactive imagination. It was real. This was not the first or the last of the giant bird sightings.

Earlier, in January 1976, Armando Grimaldo of Raymondville, Texas reported an encounter with a giant bird sailing overhead. According to Gerhard,

> Grimaldo was smoking a cigarette on his mother-in-law's porch. His estranged wife Cristina was sleeping inside the house at the time. As Armando would later put it, 'I heard a sound like the flapping of bat like wings and a funny kind of whistling. The dogs in the neighborhood started barking. I looked around but I couldn't see nothing. I don't know why I never looked up. I guess I should have, but as I was turning to go look over the other side of the house, I felt something grab me, something with big claws. I looked back and saw it and started running. I've never been so scared in my whole life.' (21)

Grimaldo reported that he could feel his attacker tearing at his jacket and shirt with its clawed talons. He escaped by diving onto the ground and crawling on his belly until he found safety under sheltering bushes.

A similar story was told by Francisco Magallanes, Jr., who described an incident that occurred at around the same time. Gerhard relates the account, though he does indicate that Magallanes admitted to having been drinking at the time:

> The twenty-one year old Magallanes told police that he went into his backyard around 12:45 am to investigate a noise and noticed an unknown creature in a stooped position. According to Francisco, the black, bat-winged animal then rose to a height of six feet and pounced upon him, scratching him badly in the ensuing struggle. Magallanes somehow managed to break free and run inside his house to safety . . . Francisco described the creature as having the face of a pig, with bright, red eyes and pointy ears, long arms, stubby legs, and an eight-foot wingspan. He claimed that the monster made a hissing sound like a snake and when it was touching him, his skin would become hot. (23)

❝ THE BLACK, BAT-WINGED ANIMAL THEN ROSE TO A HEIGHT OF SIX FEET AND POUNCED UPON HIM, SCRATCHING HIM BADLY ❞

Gerhard offers several theories concerning the possible nature of the big bird sightings. Perhaps, he suggests, 'Big Bird' is a previously undiscovered species, or a species mistakenly thought to be extinct, perhaps a teratorn – a large raptor related to modern-day condors and vultures and thought extinct for the last 6,000 years. He notes that the largest teratorn species was the *Argentivas magnificens* from South America, with a twenty-seven-foot wingspan (61).

Another possibility is that 'Big Bird' is the same as the Native American Thunderbird. The Thunderbird is a common feature in mythology and has been described as a giant vulture whose presence indicates that fierce storms are soon to arrive. The Thunderbird was thought to be a native of the spirit world as well as the physical, and Gerhard suggests that one explanation for 'Big Bird' sightings might find its source in the world of spirit, rather than science. He writes:

> I suppose we might open our minds to the idea that these events really are linked to the paranormal in some way. We should consider the possibility that these entities are, in fact, visitors from a plane of existence beyond our comprehension, somehow stepping through doors that aren't really open. (64)

Gerhard's favored explanation for 'Big Bird,' however, is that Big Bird is some sort of pterosaur, a prehistoric flying dinosaur (66). This would match the descriptions offered by some witnesses, which include bat-like wings, scaly skin, and a long tail. Gerhard theorizes that a 'group of these animals could populate remote parts of Mexico's unexplored mountains, or the marshes and jungles of Central America . . . For reasons unknown these animals may periodically migrate or roam' (68).

Researcher Jonathan Whitcomb also sees the pterosaur as a likely explanation for Big Bird sightings and sees a religious significance for the contemporary existence of physical pterosaurs. Whitcomb argues that dinosaur and pterosaur fossils have long been used as evidence of what he calls the General Theory of Evolution. According to Whitcomb, the General Theory of Evolution posits that primitive creatures such as pterosaurs have become extinct because they have changed into other forms. 'One reason,' he writes, 'people believe in G.T.E. is because of continual one-sided declarations that dinosaurs and pterosaurs are examples of primitive life that is forever gone' (39). Whitcomb began research in Papua New Guinea, site of many Big Bird sightings, to find evidence for the present-day existence of pterosaurs. Such evidence, he argues, should demonstrate that G.T.E. is not a proven fact. He writes, 'The existence of living pterosaurs is more harmonious to a belief in a Creative God than it is an accidental evolution of microbes into all the life now on this planet' (40).

Whitcomb began his search for New Guinean pterosaurs by documenting eyewitness accounts of sightings. One sighting, by a school teacher named Eunice, described an attempted grave robbery by one of the creatures, known by the natives as a 'Ropen.'

One night, in April of 1993, near the northwest coast of Umboi Island, after a large funeral procession arrived at the burial location, a creature with a glowing red tail came from the sea. (The tail was described like the glow of burning embers.) About two hundred mourners were awake when the creature flew overhead. The villagers banged pots and yelled, whereupon the intruder flew into a nearby swamp and the light disappeared. (17–18)

The creature is not only seen in flight. For example, two men reported seeing the ropen clinging to the side of a tree. The creature held itself in an upright position, looking almost like a boy climbing up a coconut tree (20).

Though not having seen the ropen himself, Whitcomb puts together a description based on the testimony of a large number of witnesses. According to his research, the creature has a wingspan of 15 meters, with an inflexible tail some seven or eight meters long. The creature is featherless, has a large head crest, and is bioluminescent, glowing eerily as it sails through the night sky. According to Whitcomb, the conclusion is unmistakable. 'Eyewitness evidence indicates a giant Rhamphorhynchoid pterosaur has an established presence in the Southwest Pacific' (132).

Pterosaurs may not be the only prehistoric lizards roaming the earth.

MOKELE-MBEMBE AND THE LOCH NESS MONSTER

According to Coleman and Clark, a 1913 German expedition to the Congo encountered a group of pygmies that informed them of an animal called Mokele-Mbembe or 'one who stops the flow of rivers.' The creature was described as the size of an elephant or hippopotamus, with an extremely long, flexible neck. It was also said to have a long tail, like an alligator. Coleman and Clark note that this description fits with the existence of sauropods or other small dinosaurs in the jungles of Africa (167).

Mokele-Mbembe is not just the subject of folklore, however. Coleman and Clark also report that in 1992 a Japanese film crew captured images of the creature on video. First noticing a large shape moving across the surface of the lake, the videographer zoomed in for a closer look. The resulting image is inconclusive but tantalizing. They write,

> The resulting footage, though jumpy and indistinct, shows a vertical protuberance at the front of the object – possibly a long neck. A second, shorter projection could be a humped back or a tail. If the object is not a dinosaur, it's difficult to say what animal it could be. (169)

The most famous candidate for a surviving aquatic dinosaur is surely the monster said to inhabit Scotland's Loch Ness. The Loch Ness Monster, affectionately known as 'Nessie,' has been the subject of press reports since 1933. On May 2 of that year the *Inverness Courier* ran a story entitled 'Strange Spectacle of Loch Ness: What Was It?' that firmly established the creature's presence in popular culture. Though this initial story indicates that Loch Ness had long been believed to be the home of a water-kelpie or water-horse (a feature shared by most large bodies of water in Scotland), it was this account that marked the beginning of the modern fascination with Nessie:

> On Friday of last week, a well-known business man, who lives near Inverness, and his wife (a University graduate), when motoring along the north shore of the loch, not far from the Abriachan Pier, were startled to see a tremendous upheaval on the Loch, which, previously, had been as calm as the proverbial mill-pond. The lady was the first to notice the disturbance, which occurred fully three-quarters of a mile from the shore, and it was her sudden cries to stop that drew her husband's attention to the water.
>
> There, the creature disported itself, rolling and plunging for fully a minute, its body resembling that of a whale, and the water cascading and churning like a simmering cauldron. Soon, however, it disappeared in a boiling mass of foam. (Binns, 10)

In August of that year, the story grew even more interesting with the publication of another eyewitness account:

> I saw the nearest approach to a dragon or pre-historic animal that I have ever seen in my life. It crossed my road about fifty-yards ahead and appeared to be carrying a small lamb or animal of some kind.
>
> It seemed to have a long neck which moved up and down in the manner of a scenic railway, and the body was fairly big, with a high back; but if there were any feet they must have been of the web kind, and as for a tail I cannot say, as it moved so rapidly, and when we got to the spot it had probably disappeared into the loch. Length from six feet to eight feet and very ugly. (19–20)

In April of 1934, newspapers published the first photograph of the creature, a photograph that came to be known as the 'surgeon's photograph.' This photograph was taken by Lieutenant Colonel Robert Kenneth Wilson, a London gynecologist, and shows what appears to be the head and neck of the creature (96). Though there have been many photographs taken of something purported to be the creature, the surgeon's photograph was for many years regarded as the best evidence, although it has recently been questioned because of claims of a hoax perpetrated with the use of a miniature model of the creature attached to a child's toy submarine.

7. The famous 'surgeon's photograph' of the Loch Ness Monster, or 'Nessie,' from 1934.

Of course, the Loch Ness Monster is hardly the only water monster to have been reported throughout the years. This category includes classic sea serpents as well as the monster of North America's Lake Champlain ('Champ' for short). There have also been countless theories as to what Nessie, Champ, and all the others might be. A.C. Oudemans' 1892 classic, *The Great Sea Serpent*, published without reference to either Nessie or Champ, suggested that sea and lake monsters might be examples of giant pinnipedia – aquatic carnivores such as seals, walruses, and otters. Long out of favor, Oudemans' theory is usually rejected in favor of the theory that creatures such as Nessie and Champ are actually extant versions of aquatic dinosaurs such as the plesiosaur.

Loren Coleman and Patrick Huyghe note, however, that eyewitness accounts would indicate that there is not simply one type of aquatic mystery creature, but many. In their *Field Guide to Lake Monsters, Sea Serpents, and Other Mystery Denizens of the Deep* they categorize these mystery creatures according to physical characteristics, range, history, likely candidates for their identity, and descriptive incidents. One of these categories is what they call the Classic Sea Serpent:

> This serpentine creature of the seas is generally quite long, upwards of 100 feet, with rough skin, a distinct head, and a tapering tail. A string of what appear to be dorsal humps, a slender neck of medium length, and noticeable eyes characterize this marine cryptid. (49)

The authors suggest that a candidate for the identification of this creature might be a zeuglodon, or ancient whale.

This Classic Sea Serpent is to be distinguished from the water-horse:

> The Waterhorse is a gigantic freshwater and marine cryptid. It has an elongated body and neck, with a rounded body showing above the water, and two sets of flippers (with the rear ones together frequently giving the appearance of a tail). The animal often exhibits a mane and, when seen at close range, seems to be covered with hair. The Waterhorse seems to have rather poor outside vision but an acute sense of hearing. (72–3)

The water-horse, unlike the Classic Sea Serpent, is found in many different environments, both marine and freshwater. These creatures are usually residents of what Coleman and Huyghe call the 'Monster Latitudes:' they live in the ocean and in freshwater lakes and rivers 'at latitudes near isothermic lines 50 degrees F, between 32 degrees F and 67 degrees F, especially in the Northern and perhaps the Southern hemispheres' (74). They argue that the water-horse is probably a type of plesiosaur.

Other types of lake or sea monsters include mystery cetaceans, giant sharks and mantas, and giant sea centipedes as well as my personal favorite, the Giant Beaver.

❝ IT HAD HUGE FRONT TEETH AND LARGE BACK LEGS AND WAS NOT . . . A NORMAL BEAVER ❞

The Giant Beaver of British Columbia is some 14 feet long with a long beaver-like tail and beaver-like legs. Coleman and Huyghe theorize that the best candidate for the Giant Beaver may be the:

> supposedly extinct giant beaver, *castoroides ohioensis*, which was the size of a black bear and weighed 600 to 700 pounds. *Castoroides ohioensis* was almost 8 feet long and had enormous, convex incisor teeth 6 inches long, 4 inches of which extended beyond the gum line. Its tail was long and beaver-like but thinner, and some paleozoologists sense that it was more round than flat. (199)

They recount a recent eyewitness report from Lake Powell, Utah. The witness reported to Coleman that:

> This beaver was about the size of a medium-sized horse or a large bear. It had huge front teeth and large back legs and was not, in our opinion, a normal beaver. It definitely had beaver-like characteristics but it was more of a prehistoric type of animal that was so large that we were all in shock. (201)

CRYPTOZOOLOGY AS SCIENCE

So what are we to make of these reports of creatures – the ones that Hawk warned me about and all the others: Wendigo, Skinwalkers, Wolf-men, Chupacabras, Mothmen, Big Birds, ropens, dinosaurs, and sea serpents? In many, if not all, of the cases the ontological nature of the creatures has seemed to vacillate between flesh and blood and 'other dimensional,' between the physical and the metaphysical.

Many researchers insist on treating these creatures entirely as empirical entities. The research into their nature is thus seen as a subset of zoology. It is a branch of zoology that focuses on unknown animals. Hence the term 'cryptozoology.' According to Coleman and Clark this term first appeared in 1959, coined by Lucien Blancou in his dedication of a book to Bernard Heuvelmans, 'master of cryptozoology' (15). Heuvelmans' book, *On the Track of Unknown Animals*, was published in 1954 and was indeed one of the first attempts to examine mysterious creatures in a scientific manner. Soon Heuvelmans and others were referring to the study of Sasquatch, Nessie, and all the rest as cryptozoology, and they were identifying the creatures themselves as 'cryptids.' Cryptozoology is thus the study of hidden animals, the study of cryptids.

In 1982 the International Society of Cryptozoology was founded and the subject matter of cryptozoology was expanded to include the study of known animals in unexpected environments, the study of animals mistakenly thought to be extinct, and the study of animals occurring in undocumented sizes, in addition to the original study of unknown or hidden animals. Thus cryptozoology is seen as the study of such hidden creatures as Bigfoot and Nessie, as well as the study of kangaroo sightings in Texas, the study of extant dinosaurs, and the study of Giant Beavers. Of course, these categories frequently overlap, as Bigfoot, Nessie, and the Giant Beaver are frequently identified as living specimens of known species thought to be extinct.

Chad Arment proposes a slightly different definition of cryptozoology in his book *Cryptozoology*, a definition that focuses less on its subject matter and more on its method. For Arment, cryptozoology is 'a targeted search methodology for zoological discovery' (9). Cryptozoology is one method, among many, for determining the status of new or lost species – what Arment calls 'mystery animals' or 'cryptids' – that are defined as 'an enthnoknown animal that may represent a new species or a species previously considered extinct' (9). Arment calls cryptids 'ethnoknown' to indicate that that encounters with these creatures have been reported in eyewitness accounts or folklore. Some cryptids may have a solid standing within a particular group's belief system, others may be known through a 'casual sighting scribbled into an explorer's journal' (11).

As a methodology for zoological discovery, cryptozoology offers the potential for success at least as great as that offered by more traditional methods. He writes,

> Rather than relying upon chance or random collection, a zoologist may investigate reports of an animal which suggest a species unknown to science. By focusing on one potential species rather than taking a scattershot approach, the scientist develops a specific procedure to locate and obtain physical evidence of the animal. Once accomplished, morphological and genetic examinations determine whether it is in fact a new species. This methodology is used in mainstream zoology, but its potential is seldom admitted. Because of risk to reputation, as it places one in the company of 'fanatics' and 'loonies', it is rarely acknowledged as a methodology in its own right – cryptozoology. (11)

Arment is careful to exclude one type of creature that might be classified as 'ethnoknown,' namely those entities that can be classified as paranormal:

> Paranormal, folkloric entities, whether ghosts, vampires, or lycanthropes, are not cryptozoological . . . Occassionally, paranormal traits are attributed to certain cryptids . . . This is usually the unfortunate consequence of poor data analysis. (11)

Indeed, Arment argues that paranormalism is a 'sister path to skepticism' because, like skepticism, it is a response to the notion that if a cryptid does exist as a flesh-and-blood creature it would have been found already. While skepticism moves from the fact that a certain cryptid has not yet been discovered to skepticism about its reality, paranormalism moves from the same fact to the assertion that the cryptid must be paranormal in nature. Cryptozoology, according to Arment, should be careful to avoid either of these faulty conclusions.

UNBELIEVABLES

As may be seen from previous accounts, it is often hard to remove the paranormal elements from the cryptids in question. If Arment is correct and cryptozoology is a method for investigating the nature of ethnoknown creatures, then surely the testimony of eyewitnesses concerning the paranormal aspects of those creatures must be taken into account. Or so believe such researchers as John Keel. Keel (2002) writes, in a style reminiscent of his hero Charles Fort,

> An almost infinite variety of known and unknown creatures thrive on this mudball and appear regularly year after year, century after century. Uncounted millions of people have been terrified by their unexpected appearances in isolated forests, deserted highways, and even in the quiet back streets of heavily populated cities. Whole counties have been seized by 'monster mania,' with every available man joining armed posses to beat the bushes in the search for the unbelievable somethings that have killed herds of cows and slaughtered dogs and horses. (1–2)

Keel notes that creatures and strange events tend to recur in the same areas, year after year. Certain geographic locations seem to have an abundance of monster, UFO, and paranormal sightings. Perhaps, he argues, this is because these locations are a type of window into another world. Eschewing the type of empirical explanation that Arment insists on, Keel argues that:

> Mundane theories do not seem to fit the known facts. We have to stretch our minds a bit and extend our imaginations into the paranormal. The sudden appearances and disappearances of these wild, unknown creatures all over the world, even in densely populated areas, suggests that they have some means of transportation or else they are deliberately dumped here and retrieved by some form of transportation. (9)

He suggests that, 'Another world exists outside our space-time continuum and that these myriad objects and creatures have found doors from their world to ours in these "window" sectors. Admittedly, it is a far-fetched idea, yet much of the data supports it' (9). Indeed, Keel suggests that mystery animals constitute at least two distinct types. Group 1:

> are genuine animals of land and sea but still unknown to science. They include at least three (probably more) types of Abominable Snow Persons, and at least seven large amphibian mammals and reptiles. Overall, they seem to be a harmless lot. They avoid us and prefer that we leave them alone. (317)

Group 2 is decidedly different: 'They are the phantoms that come crashing out of the bushes late at night' (317). About the denizens of Group 2, he writes:

> There are entities on this planet, and around it, that are far beyond all efforts to translate them into understandable cellular creatures. They are not real in the sense that we are animals motivated by sex and emotions. They are part of the energies that were scattered into space billions of years ago. Their intelligence is so vast and so ruthlessly inhuman that there is no way for us to comprehend it or communicate with it as we talk to dolphins . . . (322)

Indeed, encounters with these beings are fairly common; they may happen to anyone:

> Someone within two hundred miles of your home, no matter where you live on this earth, has had a direct, often terrifying, personal confrontation with a shape-shifting Unbelievable. Our world has always been occupied by these things. We are just passing through. Belief or disbelief will come to you from another direction.
>
> Next week, next month, or next year you may be driving along a deserted country road late at night and as you round a bend you will suddenly see . . . (327–8)

*   *   *

As the sun begins to rise over the lake, I hear the sound of a speedboat in the distance. Unable to get theirs running again, the command post has called upon one of the rangers to come to our rescue. Unfortunately the boat is pretty small and he will have to make several trips across the lake to get us all back safely. I don't wait for someone to say it is my turn – I push my way through to the front of the line and climb aboard. In minutes I am back across the lake.

When I open the door of my truck I see that there are two men sleeping inside. The storm had blown their tents away in the night and they sought shelter in my rented vehicle. This is fine with me as long as they get out quickly so that I can get back to Paris for breakfast and a shower. They are in no rush.

One of them is in a terrible mood. He is telling me about how badly he has been mistreated on this trip. He is fed up and ready to get out of Paris and back to his home.

'Hell!' he says, 'anyone can tell those tracks were fakes. Didn't they think we would notice that all the tracks were of the right foot!'

*   *   *

Pandora's box – that is what it is. Once you open the box, all kinds of things escape. Once you open that closet door, you never know what kind of monster is going to jump out. Hairy ape-men are suddenly the least of your worries. All sorts of worse things quickly come tumbling out after them – Wendigo, Skinwalkers, manwolves with fangs bared, flying Goat Suckers and winged demons, dinosaurs of the land and of the sea. It is a monstrous menagerie, a post-modern bestiary, a cantina scene of cryptids. This is what happens when you open the door.

Because the science of cryptozoology looks to folklore and eyewitness accounts for evidence of a thing's existence, its ontology is bound to grow overpopulated quickly. When you count tracks and photographs as the highest form of evidence, then almost anything can be real. If the fossil record marks the limits of what is possible, then dinosaurs can walk the earth, swim in the seas, and sail through the heavens.

Hawk was right – if Bigfoot can live in the north Texas woods, then so can Wendigo, so can Chupacabras, so can Mokele-Mbembe. If cryptozoology is the study of ethnoknown creatures, then cryptozoology has its hands full. Fantasy will always be more heavily populated than reality and Keel's Unbelievables will always haunt the borderlands.

I, for one, prefer overpopulation to scarcity, diversity to monotony. There is enough talk of extinction among the creatures whose existence we all agree on, polar bears going the way of the polar ice. It seems that humanity has a fine record of disposing of creatures just as quickly as we make their acquaintance. It is nice to hear that there may yet be mystery in the world, that there may yet be things that we have not yet classified and put into a box – things that, by refusing to be known, are refusing to be destroyed.

Not that I am willing to say that all of the creatures discussed in this chapter are flesh-and-blood animals. My personal ontology tends to be quite a bit smaller than that. (Besides, I don't know if anyone believes in the existence of all these things – surely even the most devoted cryptozoologist has to pick and choose, perhaps taking Bigfoot and Nessie, and leaving Chupacabras behind.) This does not mean that I am bothered that other people do believe these things. I tend to prefer diversity in this case as well. After all, just because I can't believe in something doesn't mean that I think poorly of those who can. (In some ways, I suppose, I admire them.) It has always seemed to me that the human race needs more things to wonder about, rather than less. It is also rather refreshing to think that the publishers of textbooks and the curators of museums don't have the last word about what is real and what is not. There is nothing like a good, old-fashioned democracy of ideas to keep things interesting. This means that we will have to tolerate a great many weird notions and bizarre beliefs. Of course, for some of us, it is less a matter of tolerance, more a matter of enjoyment. The more the merrier, I say.

And if the world does happen to include a one-legged, tree-climbing, rock-throwing, marsupial Sasquatch that prowls the wilds around Paris, Texas, that will be all right by me.

THREE

The Spirit of Sasquatch

I am still looking for Bigfoot, though I have fled Texas to do so, riding through the bottom lands of southern Arkansas in search of the small community of Fouke and the legendary monster of Boggy Creek. For me this is a trip back in time, back to 1972 and the Hatfield Roller Rink and Drive-In Theater where I first watched The Legend of Boggy Creek. *The low-budget film purported to be a true account of events that had taken place around Fouke, Arkansas, events involving a large, three-toed, hairy hominid called the Fouke Monster. More than one Bigfoot believer that I have talked with has traced the genesis of their fascination with the creature to this film.*

There is not much left of the monster in Fouke these days: a statue of the beast stands on the side of the main highway, with a cut-out face made for photographing yourself in the guise of the creature; a small general store has a dusty display of items related to the movie, including tee-shirts emblazoned with a drawing of the creature and the words 'The Legend Lives;' and, just off the main road, Smokey Crabtree, star of the film and the source of many of its stories, operates 'Smokey's Two Books Bookstore' where he sells, among other things, copies of his three books. Smokey also claims to have the carcass of the creature, which he will show to select visitors.

I contacted Smokey by telephone to request an interview. He turned me down in no uncertain terms.

'Who the hell gave you my phone number?' Smokey demanded.

I informed him that I had found the number on his website.

I asked if he would agree to an interview if I came to visit in Fouke.

He said something about being screwed by people like me so many times that he always kept a jar of Vaseline handy.

'Could I at least stop in to see the creature that you have on display?' I asked.

'Who the hell told you about that?' he shouted.

8. The author as the Fouke Monster, Fouke, Arkansas.
Only not as good looking.

'*I read about it on your website.*'
'*If you want to know my story you can come and buy my books.*'
   So I am on my way to buy his books. My traveling companions are my wife, Kristen, and my friends, Sandy, Cheryl, and Michael. Smokey lives in a modest brick house. A mobile home is parked out front. On the mobile home is a large sign announcing 'Smokey's Two Books Bookstore, Fouke Monster Souvenirs, and Used Books Museum.' He greets us in the yard with a big smile and a firm handshake. Apparently, he doesn't recognize me.
   The bookstore sells not only Smokey's books but a whole host of used books and a nice selection of Fouke Monster memorabilia. The walls are lined with mementos from Smokey's life and with newspaper and magazine articles about the monster and the movie. I have already purchased tee-shirts at the general store, but I pick up a shot glass, a CD entitled The Legend of

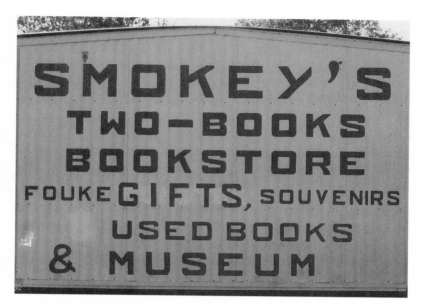

9. Smokey's Two-Books Bookstore, Fouke, Arkansas.

Smokey Crabtree, *and Smokey's three books. My friends also buy a lot of stuff. We are hoping that our purchase will earn us a glimpse of the creature. We are not disappointed.*

THE LEGEND OF BOGGY CREEK

The Legend of Boggy Creek is, hands down, the greatest Bigfoot movie ever made. (Yes, I am including *The Geek*, that 1970s bit of Bigfoot pornography, when I make this assessment.) Its mixture of documentary-style story-telling, location shots in the swamps of Arkansas, and Arkansans re-enacting their encounters with the creature somehow manages to overcome the cheap gorilla suit and cheesy songs about the creature's longing for love. A drive-in movie hit when it was released, it is now a cult classic among horror movie buffs and Bigfoot believers. The film tells the story of the Fouke Monster's visitations to the area around Fouke and Boggy Creek. The creature leaves its three-toed tracks in bean fields, kills hogs, and terrorizes teenagers and young married couples, but never does too much harm. Perhaps it is not the greatest film ever made but it scared me when I first saw it as a kid and, I admit, it scares me a bit even today.

Smokey Crabtree has a less favorable impression of the film, however, and doesn't mind telling the world about it in his first book *Smokey and the Fouke Monster*. The book combines folksy stories about growing up poor in Arkansas with Smokey's account of how the movie got the story wrong and how the people of Fouke were taken advantage of by unscrupulous movie people. It also includes Smokey's account of his and his family's encounter with the monster.

Smokey's son Lynn had one of the closest encounters with the creature while squirrel hunting with his dog. Hearing the dog 'hollering out in pain like he was hung in a fence,' Lynn ran to investigate and help the dog out of the barbed wire. Before he reached the fence, however, the sound changed into a groan or grunt. Lynn realized that the sound wasn't coming from the dog. It was a sound like he had never heard before.

He eased out to an opening so he could see.

> There, only thirty steps away, was a hairy animal of some kind. He had his back to Lynn . . . He stood seven or eight feet high, had reddish brown hair, about four inches long, completely covering his body. He was standing up in the form of a man or gorilla. His arms were extra long. Lynn told me later that he was standing real odd, like a man would stand if he was using the stool standing up . . . (84–5)

Lynn waved the gun around, thinking he might scare the creature away. Instead, it started walking toward him.

Lynn shot him in the face hoping to damage his eyesight:

❝ THE THING NEVER FLINCHED . . . LYNN RAN BACK A FEW STEPS AND SHOT HIM TWICE MORE ❞

The thing never flinched or changed his stride of walking. Lynn ran back a few steps and shot him twice more, emptying his gun that time. He was trotting along ahead of the creature reloading his gun. He said by the time he had finished loading the gun he was scared and running. The farther he ran the more scared he became. When he got to the house he was completely panicked. He was in a state of shock. (84–5)

In his latest book, *The Man Behind the Legend,* Smokey describes a skeleton found by hunters in the early 1990s. The hunters, unsure of what they had found, donated the skeleton to Smokey because of his connection to the Fouke Monster story. Smokey seems convinced that the skeleton represents the remains of whatever it was that terrorized the people of Fouke for so long:

> The eight-foot skeleton that was found in the woods in our part of the country, and that I have had in my possession for over ten years now, has been the most convincing evidence of the creature's existence that we have been able to find. The skeleton is of some specimen that was over eight feet tall and weighed in the neighborhood of six hundred pounds when it was alive. It's very obvious that the skeleton is from a creature that should not be, and does not belong, in our environment in this part of the country. Through research, we are convinced that when it was alive, it was capable of doing at least ninety-five percent of everything that has happened around this part of the country, including the day time sightings of this creature that were experienced by good, honest, civilized people in the community. The skeleton is priceless to me, regardless of what it is, because it comes closest to solving my personal mystery and clearing up my mind. (48)

Not everyone is so sure, of course. But then, despite its influence on so many Bigfoot believers, the Fouke Monster story has never been accepted in mainstream Bigfoot research. Perhaps it is simply because the Fouke Monster is a resident of southern Arkansas rather than the Pacific Northwest. Perhaps it is because the tracks of the creature show three toes rather than Bigfoot's five, or because details of sightings are somewhat confusing, or that the story burst on the scene in movie form before any of the more famous Bigfoot researchers had a chance to investigate them. Smokey's strong sense of ownership of the story and of the evidentiary carcass also complicates matters, as does his reluctance to allow the Fouke Monster to be subsumed under the larger category of Bigfoot or Sasquatch, rightly noting that the Fouke Monster story had a life of its own before 'Bigfoot' became a cultural phenomenon. Consequently, the Fouke Monster is strangely outside mainstream Bigfoot research, Smokey's skeleton is something of a skeleton in the closet. It is not the only skeleton, however.

SEX AND THE SINGLE SASQUATCH

A troubling feature of many Bigfoot accounts is the claim to have had experiences with Bigfoot creatures over long periods of time – periods in which the Bigfoot seem to become habituated to human presence. One of the better-known accounts is that of the pseudonymous Jan Klement in the book *The Creature*:

> Sometimes the mind plays strange tricks upon us and with the passage of time we find it difficult to separate truth from fantasy. Time has passed, about two years to be exact, and I feel that I better write down my story before it passes into that gray area of unreality.
>
> You who read this story can expect no great prose, no witticisms to tempt the chuckle bones, nothing to challenge the imagination. This will be a straightforward account of what happened. You may believe it if you wish or refuse to believe it. Your acceptance or rejection of this tale will have no effect on the future of the world or the future of yourself. (5)

Thus begins the account of one man's experience with a Bigfoot in southwestern Pennsylvania, a tale of friendship between a human and a Bigfoot creature, affectionately named 'Kong.'

Having built a small cabin on 11 acres of land, Klement worked hard to improve the property, which he called 'The Diggins' because of his inclination to dig ditches, holes, ponds, and embankments on the land. On one occasion, after working all day to dig a small pond, Klement settled onto the porch of the cabin to enjoy a beer:

> I heard a slight noise to my right and so I lifted myself and peered around to the side of the cabin from the porch and there crouched before me was a large hairy creature. In an instance the creature turned and leaped into the brush to the back of the cabin and was gone. I stood stunned. 'My God.' I must have shouted it out loud. (7)

A few days later the creature returned. This time Klement was eating an apple on the cabin porch when the creature suddenly appeared at the porch railing. The creature carefully extended its arm, snatched two apples, stared briefly at Klement, and then ran away into the bushes. Klement estimated the creature was over seven feet tall, man-like or ape-like in appearance, and covered with short brown hair. The creature had large eyes and an 'expressive' mouth. Klement noticed well-developed leg and shoulder muscles and a protruding stomach. The stomach warrants an explanation from Klement:

> The protruding stomach may seem normal when one considers apes and sees pictures of them but when an artist conceives a human the human is pictured as flat stomached and muscular which is a misconception. Stand on any street corner and you will observe that humans are flabby with protruding guts. The older the human the more protruding the gut. I can say this, as I look down over my own creeping obesity. So you should not consider the creature as too much different than humans on that basis. (8–9)

Following this second meeting, Klement actively attempted to attract the creature and to habituate it to his presence. Through trial and error, he discovered that the secret was to work until his clothes were damp with sweat and then to sit on his porch with apples for bait. When the creature appeared again, Klement encouraged it to come closer by hand gestures. This seemed to have no effect. Klement then tossed an apple at the creature's feet. It was then that he saw that it was a male:

> Its penis hanging limply in front, scarcely noticeable in the failing light. He picked up the apple, opened his mouth and with one chomp crushed it in powerful jaws. I threw him another and another and on the third apple I spoke in low hushed tones increasing my voice to normal about the tenth apple. I was now fishing apples from the chip basket on the railing. The creature loaded most of these in his arms and walked to the bushes at the end of the small clearing and disappeared. (10)

The next evening the creature returned and this time would take apples directly from Klement's hand. Klement notes that the creature most certainly did not subsist on apple hand-outs. Indeed on one occasion he saw it run down a small deer and carry it away into the bush, presumably for a feast.

Klement describes why he was not frightened by these events:

> I have always prided myself on not being afraid of anything and this seemed to be a situation in which the creature was afraid and I was the aggressor. I have traveled somewhat in Europe and have slept among hostile gypsies in Hungary and faced a knife point in an alley off Piccadilly Circus in London. When these adventures occurred to me they seemed unreal and as if it were happening to someone else,

perhaps to my alter ego or some other psychological manifestation of my personality. It was and is this sense of unreality about life that makes me unafraid. Besides I was a scientist and this was a situation in which all true scientists wish to find themselves someday. (11)

In a few weeks' time, Klement had managed to tame the creature somewhat and gave him the name Kong. He even managed to teach him a few simple words. He taught the creature the word 'Stay' and, perhaps more importantly, 'Go' – a word used to tell the creature to leave the area quickly.

> The word and sound *BAH* meant *NO*. I tried to teach him *NO* but he confused it with *GO* and would leave the area. I usually had to use *BAH* when he would touch me too briskly or start to eat one of the small surviving spruce trees I wished to save. (35)

Over time Klement found it hard to think of Kong as an animal. In addition to his physical resemblance to a human being, Klement noted that Kong also seemed to share emotions in common with our species. Facial expressions indicated pleasure, anxiety, fear, surprise, and annoyance: 'Once I thought he was smiling but it was only a grimace, probably from gas which he often expelled orally and anally with explosive force' (36).

On one occasion, Klement even observed that Kong was sexually aroused:

> Well here was Kong with a glowing hard on, standing in front of me. I gave him two apples which he promptly ate. As he stood around I felt uneasy and once again embarrassed. Occasionally he would touch one end of his penis and seemed to be brushing away flies but none were in evidence. He was not masturbating. Finally I told him to get the hell out of here and if he had a female to go and find her. Of course he didn't understand what I was saying . . . (65)

Kong did find a way to relieve the sexual tension, however, much to Klement's surprise. Looking across a field at a herd of cattle, Klement saw Kong in action:

> He was mounted on a large Holstein cow and was shoving away. The cow would start to walk away and Kong would lift his legs and hang on with his hands cupped against the side of the cow until it would stop and then he would begin working his buttocks rapidly again . . .

I told Kong that there were no more apples and that he should go. I headed for the car and Kong started slowly up the road toward the top of the hill. I hollered after him 'you picked the ugliest one.' (66–7)

Before long, however, Klement noticed that Kong seemed to be ill, perhaps even feverish. One day he returned to the cabin to find Kong lying in the rain, dead. Fearing the public's intrusion into his privacy, Klement decided to bury Kong secretly. With a great deal of effort, he managed to pull Kong into the back of his station wagon with the aid of a rope and an old door for a ramp. Once inside the car, he drove Kong to a deserted spot many miles away for the burial. Getting Kong out of the station wagon and into the woods proved to be quite difficult, however, and the whole affair ended in a rather grisly scene:

I tried to drag Kong into the woods but he was too heavy. There was only one thing to do. I took out the axe and started cutting him into movable pieces. First his head came off with one deft stroke. There was a clap of thunder and it started to rain. My feet were in snow and my head was in rain. His arms came off with difficulty. His legs needed several hacks and by this time I was crying uncontrollably. Some blood formed on the wet snow. I took out my bottle of scotch and swilled much of it down. Then I started carrying the pieces into the woods. I have no idea how far I went but it was a good ways. I made one trip for the arms, one each for the legs. And one for the head. I had to drag the torso behind me with the rope. It kept getting tangled on greenbrier and small saplings.

The rain pelted me, the lightning flashed and I could see West Virginia over on the other hillside as I dug the grave. The pieces of Kong lay in disarray about me. Finally the hole was about three feet deep and full of water. I dug a slit trench in the hillside to drain most of the water away. Snow was still in abundance on the ground.

The covering job was a nightmare and I was glad when it was completed. After the job I repeated 'I am the resurrection and the life sayeth the Lord and he who believeth in me though he were dead yet shall he live and he whoever liveth and believeth in me shall have everlasting life.' I don't know why I said it, I'm not a religious person but this laying to rest of one of God's creatures evoked this testimony of sorts from me. I returned to the car with my shovel, my rope, and my poncho. When I got there I finished the rest of the scotch, got in the car, threw the bottle out the window, turned the car around and skidded up the hill to the gap and down the other side. (84–5)

The Locals

Thom Powell, in his book *The Locals: A Contemporary Investigation of the Bigfoot/Sasquatch Phenomenon*, provides accounts of other individuals and families who have claimed to have had personal encounters with Bigfoot over extended periods of time. Indeed, Powell believes that there are multiple instances on record that show that the Bigfoot creatures may become habituated and learn to interact with humans. He believes that these examples of habituation describe the best method of contacting and establishing proof for the existence of the creatures. Powell believes that a large tract of forested land, in any part of the U.S., should yield a Bigfoot, provided that the residents of that land properly habituate the creatures (103).

Habituation requires several elements and Powell offers pointers for habituating a Bigfoot of your own, including what he calls 'Landscaping for Sasquatch.' Your residence should be relatively secluded and should include a reasonably large section of forested land. A ready source of food should be available – an orchard of fruit trees is often successful. You should not actively seek the creatures, but go about you own business. They don't like to be sought out or watched. They also don't like cameras – one flash or shutter click might drive the creatures away for good. Be patient – it could be years before you begin to see signs of the animals' presence. Women and children will probably have the best chance of encountering the creatures, they seem to be preferred over men. Don't try to shoot a creature with a gun. Where there is one Bigfoot there are usually more and if you kill a Bigfoot you might find that you will have to defend yourself.

> **❝ SHE REMEMBERS ONE CREATURE LOOKING IN HER OPEN WINDOW AND PLACING ITS LARGE HAND ON HER CHEST ❞**

Powell quotes Grover Krantz on the issue of what to do if you manage to shoot a Bigfoot – reload!

One of several accounts that Powell relates is that told by Dora Bradley. Growing up in rural Missouri in the 1960s, Dora was not aware that there was anything strange about encounters with Bigfoot. She remembers one creature looking in her open window at night and placing its large hand on her chest. She also relates a time when she, her brother, and a friend played with a juvenile Bigfoot. The young animal played rough and smelled terrible and Dora was not keen on the idea of playing with it again. Once, she awakened to discover that she and her brother had been taken by the creatures to a small cave. Surrounded by the creatures she saw one of them digging a hole to bury what looked like an infant: 'The bigfoot sitting next to me pulled me closer. It put my head on its chest. I felt its breasts when I pushed its chest. I thought it tried to nurse me the way my mother breastfed my baby sister' (146). Soon, another of the creatures lifted her in its arm and took her back to her family's garden.

Powell says that he chooses to think of the creatures as 'The Locals:'

> They have long been a part of the landscape that we have recently adopted for our human purposes. Like the Native American, they were here before we arrived. Unlike the Native Americans, we have not succeeded in evicting them. They do not fear us; in fact, they observe our human activities with great interest. They stay in the shadows most of the time, though they occasionally manifest themselves. When they do, a few people notice. Most do not. (8)

And again, 'We're the out-of-towners. We've just arrived on this continent. There's a whole lot we could learn from "the locals"' (257).

*　　*　　*

I am listening to sound recordings of Bigfoot in the wild. First there is loud tree knocking.

Whump! Whump! Whump!

Then vocalizations – high-pitched hoo! hoo! hoo!

This is followed by voices. At first it sounds like an LP record played backwards, the way we used to do when I was a kid. We were trying to find subliminal satanic messages that rock bands had hidden in their songs, a process called 'backward masking.' Just like with backward LPs, I can make nothing out, except that there seems to be more than one individual making the sounds.

Fortunately I have a translation of the sounds, which are spoken English. At first the words on paper do not seem to match what I am hearing. Then that changes. I can hear them clearly now, at least as long as I read along on the transcript.

A high-pitched voice, identified as a female, says, 'Who, who, who, who, who, who.'

Then a deeper, male voice says 'Ronock told them what a good boy I am though.'

The female again: 'Who are you all? Who are you all? They're so funny looking. Watch this. Who are you all?'

A human voice answers back in imitation.

The female asks 'Why's he doing that?'

The male responds 'Oh, because he put the paper down. Good one.'

The female asks again 'Who are you all?'

The male says 'Oh Ronock told me what a good boy I am.'

The female: 'Who are you all? Want good will? Ya want to? Nawk. Want Nawk? Nawk?'

The human voice answers back and the female responds 'What's he doing?'

Okay, I understand most of the words now, but I can make no sense of them.

I am reminded of Wittgenstein's remark that 'If a lion could talk we could not understand him.'

FOREST FRIENDS

Janice Carter Coy is the translator of Bigfoot speak (www.bigfoot-referenceguide.com). She claims to have developed her ability to understand the creatures through what she describes as a lifetime spent among them. Now an adult, Carter Coy first encountered a Bigfoot when she was 7 years old. Playing on her grandfather's farm in rural Tennessee, she literally ran into one of the creatures. Frozen in fear, Janice was rescued by her grandfather who rushed to her side and proceeded to stare the creature down until it moved away.

A year or so later, her grandfather took her out to the top of a nearby hill to show her the creature again. He had brought table scraps to feed it. He sat the plate on the ground and backed away while it came forward and ate. As time went on Carter Coy came to realize that there were several of the creatures, whom she calls 'Forest Friends,' living on the farm. Her grandfather had first befriended them in the 1940s when he rescued an injured juvenile and nursed him back to health. He named the creature 'Fox.' When her grandfather became ill and could no longer go out, the creature would come and sleep under the family mobile home, sometimes destroying the underpinning and heating ducts. When her grandfather passed away, Janice Carter Coy took up the responsibility for feeding and interacting with the creatures. In recent years Janice decided to tell the family story, with the help of Mary Green, in the book *Fifty Years with Bigfoot: Tennessee*

Chronicles of Co-existence. Green claims that because she grew up among the Forest Friends, she has developed a lasting relationship with them. She also claims that, in addition to their own language, the creatures have learned a good deal of English from the Carter family.

Russian Bigfoot researcher Dmitri Bayanov has focused on the Janice Carter Coy case, especially the claim that the creatures can speak English, in his argument that Bigfoot are more than animals, if not quite human. They are, he suggests, best considered to be 'manimals.' The fact that Carter Coy's grandfather could teach young Fox English and could co-exist in a complex social relationship with the Bigfoot indicates that the creatures are closer to humans than are any other known species.

Also of importance to Bayanov is Carter Coy's story of a burial of a stillborn Bigfoot baby. Digging a hole with their bare hands and with pointed sticks, the creatures buried the infant. For a long while after the burial, the mother would deliver food to the grave, as if expecting the baby to eat. This suggests, hints Bayanov, that the Bigfoot may even have a concept of the after-life.

Perhaps most important of all, however, is the claim that the Bigfoot on the Carter farm possessed a language of their own. It is not only that they acquired a kind of rudimentary English but that they possessed a spoken language independent of their English training. Carter Coy learned the basic of Bigfoot language from her grandfather and from Fox and others who could speak both Bigfoot and English.

Bayanov offers an example from Carter Coy as evidence of the depth and complexity of Bigfoot language. As a child Fox scared Janice and her little sister Lila. Her grandfather quickly chastized Fox for frightening the girls. By way of apology Fox spoke in the Bigfoot language to Lila and said 'Yoohhobt Papi Icantewaste Mitanski . . . Posa . . . Ka Taikay Kataikay Tohobt Wabittub.' This is translated into English as 'Yellow Hair, be happy little sister. I naughty. Don't cry Blue Eyes.' Bayanov writes:

> In other words, Fox was apologetic, tried to console little Lila and used her traits in naming and addressing her. All that in a few touching words. Call him what you like: bipedal ape, Australopithecus robustus, Gigantopithecus blacki, for me such an utterance, if it really happened, is the sure sign of a human being. (www.hominology. narod.ru)

Indeed, accounts like the one given by Janice Carter Coy are a central part of why Bayanov rejects any attempt to kill a Bigfoot for study and dissection. For Bayanov, this would be the equivalent of murder.

Janice Carter Coy's account of her family's multi-generational relationship with a family of Bigfoot represents an extreme version of the habituation championed by Powell and illustrated by Klement's tale. It also suggests that Bigfoot are a long way from the super-gorillas envisioned by Krantz and Meldrum. Not only does Carter Coy claim to have lived alongside a family of the creatures for most of her life, she also claims that they exhibit very complex social and linguistic behavior. It should not be hard to imagine why her story, and others like it, are rejected by many Bigfooters. Carter Coy's claims about the abilities of her Forest Friends are not the most extreme, however. There are plenty of others who see in Bigfoot a being that is more than a fellow resident of our planet earth, more than a physical entity, whether *Gigantopithecus* or manimal.

APE CANYON

Originally reported in 1924, Fred Beck's story of an attack by ape-creatures upon a cabin on Washington's Mount St. Helens became news again following the renewed interest in Sasquatch stories in the 1950s. The 1924 story, which caused quite a stir in regional newspapers at the time, told of how Beck and a group of prospectors had been attacked while in their cabin. The Portland, Oregon *Oregonian* of July 13, 1924 first broke the story:

FIGHT WITH BIG APES REPORTED BY MINERS
FABLED BEASTS ARE SAID TO HAVE BOMBARDED CABIN
ONE OF ANIMALS, SAID TO APPEAR LIKE HUGE GORILLA,
IS KILLED BY PARTY

Kelso, Wash., July 12 – (Special) The strangest story to come from the Cascade mountains was brought to Kelso today by Marion Smith, his son Roy Smith, Fred Beck, Gabe Lefever and John Peterson, who encountered the fabled 'mountain devils' or mountain gorillas of Mount St. Helens this week, shooting one of them and being attacked throughout the night by rock bombardments of the beasts.

The men had been prospecting a claim on the Muddy, a branch of the Lewis River about eight miles from Spirit Lake, 45 miles from Castle Rock. They declared that they saw four of the huge animals, which were about 7 feet tall, weighed about 400 pounds and walked erect. Smith and his companions declared that they had seen the tracks of the animals several times in the last six years and Indians have told of the 'mountain devils' for 60 years, but none of the animals ever has been seen before.

Smith met with one of the animals and fired at it with a revolver, he said. Thursday Fred Beck, it is said, shot one, the body falling over a precipice. That night the animals bombarded the cabin where the men were stopping with showers of rocks, many of them large ones, knocking chunks out of the log cabin, according to the prospectors. Many of the rocks fell through a hole in the roof and two of the rocks struck Beck, one of them rendering him unconscious for nearly two hours.

The animals were said to have been the appearance of huge gorillas. They are covered with long, black hair. Their ears are about four inches long and stick straight up. They have four toes, short and stubby. The tracks are 13 to 14 inches long. These tracks have been seen by forest rangers and prospectors for years.

The prospectors built a new cabin this year and it is believed it is close to a cave thought to be occupied by the animals. Mr. Smith believes he knows the location of the cave. (Green, 45)

Several follow-up articles were printed but interest soon faded away.

The story was rediscovered in 1964 by John Green and Roger Patterson, who each interviewed Beck concerning his experiences. Green noted some discrepancies in Beck's testimony, writing that: 'To my understanding there was a difficulty in fitting all the elements of his story in logical order, but I was not able to clear that up' (48). Green was not sufficiently troubled by these difficulties to discount the veracity of the account, however. He wrote:

> Did all this really happen? I think so. To the people at that time and place, knowing nothing of such creatures except the old legends of mountain devils, the miners' story was not believable. However if such animals do exist, then certainly the most acceptable explanation for the miners having claimed to see them is that they did see them. There isn't a shadow of a suggestion as to why they would make up such a story and keep telling it all their lives. (48)

Most of the early researchers would consider the story of Ape Canyon one of the classics of the field.

In recent years, however, Beck's autobiography has caused some Bigfooters to take a second look at the Ape Canyon story. In particular the account, published in 1967 as *I Fought the Apemen of Mount*

St, Helens, WA, by his son Ronald A. Beck, adds some decidedly
paranormal aspects to the story that many find troubling.

Beck's story begins with a prospecting trip to Mount St. Helens,
Washington and an overnight stay in a cabin. After discovering a
series of strange tracks, the prospectors were apprehensive, but
nevertheless decided to stay on for another day of gold hunting. As
the men settled into the cabin for dinner, they began to notice strange
whistles coming from the surrounding woods. A whistle on one side
of the cabin would be answered by a whistle from the other side,
giving them the impression that the cabin was surrounded. They
also heard strange booming or thumping sounds, like the sound of
something beating itself on the chest.

Going to the spring to get water, Beck and another prospector saw
one of the creatures and shot at it, only to have it run away down a
little canyon. As it was fast becoming dark, they made their way back
to the cabin in order to tell their bunk-mates the story. It was decided
that they would spend the night in the cabin and then head for home
as soon as the sun was up. About midnight the men were awakened
by a loud noise. Something had knocked the chinking loose from
between the logs of the cabin walls. Looking through the hole in the
wall, Beck saw at least three of the creatures:

> This was the start of the famous attack, of which so much has been
> written in Washington and Oregon papers through out the years.
> Most accounts tell of giant boulders being hurled against the cabin,
> and say some even fell through the roof, but this was not quite the
> case. There were very few large rocks around in that area. It is true
> that many smaller ones were hurled at the cabin, but they did not
> break through the roof, but hit with a bang, and rolled off. Some did
> fall through the chimney of the fireplace. Some accounts state I was
> hit in the head by a rock and knocked unconscious. This is not true.
>
> The only time we shot our guns that night was when the creatures
> were attacking our cabin. When they would quiet down for a few
> minutes, we would quit shooting. I told the rest of the party, that
> maybe if they saw we were only shooting when they attacked, they
> might realize we were only defending ourselves. We could have had
> clear shots at them through the opening left by the chinking had we
> chosen to shoot. We did shoot, however, when they climbed up on
> our roof. We shot round after round through the roof. We had to brace
> the hewed-logged door with a long pole taken from the bunk bed.
> The creatures were pushing against it and the whole door vibrated
> from the impact. We responded by firing many more rounds through

the door. They pushed against the walls of the cabin as if trying to push the cabin over, but this was pretty much an impossibility, as previously stated the cabin was a sturdy made building. (www. bigfootencounters.com)

The most frightening moment came when one of the creatures pushed his arm between the logs and into the cabin. The creature managed to grab hold of an ax, but Beck stopped the creature from taking it by turning the ax head in such a way that it would not fit between the logs. One can only imagine the kind of damage that might have been done to the cabin if the creature had managed to pilfer the ax. The attack stopped with daybreak. It was then that Beck spotted one of the creatures and fired upon it. It fell into a gorge, roughly 400 feet down. The men quickly made their way down the mountain.

❛ ONE OF THE CREATURES PUSHED HIS ARM BETWEEN THE LOGS AND . . . MANAGED TO GRAB HOLD OF AN AX ❜

After relating his account of the events of Ape Canyon, Beck then offers what he calls background information to the story. It is here that the story begins to take on the paranormal tone that has caused some investigators to discount the entire account:

> In the first chapter I told about the attack, and now I want to go into the background, and tell a little concerning our activities. They will be colorful, and from them emerge a spiritual and metaphysical understanding of the case.
>
> First of all, I hope this book does not discourage too much those interested souls who are looking and trying to solve the mystery of the abominable snowmen. If someone captured one, I would have to swallow most of the content of this book, for I am about to make a bold statement: No one will ever capture one, and no one will ever kill one – in other words, present to the world a living one in a cage, or find a dead body of one to be examined by science. I know there are stories that some have been captured but got away. So will they always get away.

Beck's certainty that the creatures will never be captured comes from his understanding that the creatures 'are not entirely of the world.' Throughout the attack, Beck insists, he was always aware that they were dealing with supernatural beings. He also insists that the other men in the cabin felt the same. Indeed, psychic methods had been used by the men to locate the mine:

> Back in 1922 we found the location of our mine. A spiritual being, a large Indian dressed in buckskin, appeared to us and talked to us. He was the picture of stateliness itself. He never told us his name, but we always called him the Great Spirit. He replied once, 'The Great Spirit is above me. We are all of the Great Spirit, if we listen when the Great Spirit talks.'
>
> There was another spiritual being which appeared to us – more in the role of a comforting friend, and we learned her name. One of our party suggested later that we name our mine after her; and so the mining claim we later filed bore her last name. The big Indian being told us there would be a white arrow go before us. Another man, who was not present during the attack in 1924, could see the arrow easily and clearly at all times. And I could see it nearly as well.

After following the white arrow for four days, things took a turn for the worse, however. One of the prospectors began to lose patience and cursed the spirit guide. When they finally reached the mine location, their spirit guide had some difficult words to share with them as a consequence:

> We got a little closer, and we all saw the image of a large door open, and the big Indian appeared in front of it. He spoke: 'Because you have cursed the spirit leading you, you will be shown where there is gold, but it is not given to you.'

The apes of Ape Canyon were lesser forms of spirit, intent on keeping the prospectors from discovering the true riches of the mine.

Beck offers as evidence for his theory of the psychic origins of the creatures a series of footprints found on a sandbar in a creek. In the center of the sandbar, which was about an acre is size, Beck and his friends discovered two large bare footprints, four inches deep. They were the only tracks:

> There we were standing in the middle of the sand bar, and not one of us could conceive any earthly thing taking steps 160 feet long. 'No human being could have made these tracks,' Hank said, 'and there's only one way they could be made, something dropped from the sky and went back up.'

There was no third step. This is certainly another indication of what I'm saying about manifestation.

Beck surmises that the Abominable Snowmen, as he calls them, come from a lower plane of reality than our own. When vibrations are at the proper frequency they may pass from one plane to another and appear as physical beings. These beings are not animal spirits, but they are also not human – they fall inbetween. They are 'the missing link in consciousness' between apes and humans. Their metaphysical nature explains why physical remains of the creatures have never been found.

Most theories picture the Snowmen as material beings hiding in caves, and scampering over the mountains. The law of probability would be that eventually one would be found if their bodies were of physical construction only. If one claims only the physical laws to explain their existence, then we can use a material logic to prove or disprove the premise. If they are material life definite material evidence would surely be found.

The mystery of the Sasquatch cannot, then, be solved through expeditions to the wood, but through human self-reflection and spiritual awakening. In order to solve the mystery we must 'break the little material shell' that surrounds us and seek a higher order in the spiritual realm. As Beck concludes:

I have lived this experience with Abominable Snowmen. I have encountered them on the slopes of Mount St. Helens. I have looked deep into myself to tell you of their nature.

I have had both the earthly experience of encountering them by Ape Canyon, and the spiritual experience of knowing and telling what they are.

I have walked through the messy cliffs of Ape Canyon, and seen a primeval loneliness, reminiscent of life as it must have been years ago.

I have explored the distant future which beckons to us with hope. I have told you my story and it is true. Abominable Snowmen are a part of the creation. Will we hear much more from them? Will their habitat change from selected mountains to nearer our populous cities?

I think they will. They are just one little mystery from the ocean of mysteries.

10. Bionic Bigfoot from space.
Andre the Giant as Bigfoot in TV's *The Six Million Dollar Man*.

Far from being an aberration among Bigfoot accounts, the story of Ape Canyon would set the pattern for many paranormal interpretations of the creature to follow.

PSYCHIC SASQUATCH

Did you know that Sasquatch can: read, write, shape-shift, voice project, create infrasound that affects the environment, de-materialize at will or cause you to have an experience of lost time so you think they de-materialized, travel 300 miles a day on foot, live in well-lighted underground facilities, contact and live with Star People, tell us about our past and our future, have lived here longer than the human race? (www.joanocean.com)

According to Joan Ocean, this is all true. Bigfoot are much more than large bipedal primates. They are much more than manimals, lying somewhere between humans and apes in development, language, and culture. They are also a lot more than the inhuman spirits of Ape Canyon. Bigfoot are our spiritual, technological, cultural, and intellectual superiors. If we let them, they will be our teachers and guides.

Ocean's interest in Bigfoot grew out of her belief that it is possible to communicate with dolphins as she described in her 1989 book *Dolphin Connection: Interdimensional Ways of Living*. The connection with Bigfoot began when she was contacted by someone asking if her techniques for communicating with dolphins might be used in communication with Sasquatch. Intrigued by the possibility, Ocean began to seek out encounters with the creatures.

Her first attempt at contact, in an undeveloped spot in the Hawaiian islands, resulted in several experiences that she only identified as Sasquatch communication at a later time. During a week-long camping trip she experienced noxious odors, the sounds of falling trees and crooning birds, and gentle taps on her shoulder while she was trying to sleep. Blind to it at the time, Ocean now believes that these were subtle attempts by the creatures to initiate contact with her.

Some time later Ocean learned that a friend who lived near a forested area believed that Sasquatch creatures lived nearby. Ocean sent a copy of her book to her friend as a gift to the creatures. In response, Ocean received a gift from the Sasquatch, a beautiful quartz rock with an attached note in printed letters that read 'FOR WATER WOMAN.' Despite her initial shock, Ocean soon learned that:

It came from the Medicine Woman of the Sasquatch family in a southeastern location of America near my friend who I will call Susan to ensure her privacy. Apparently the great grandmothers of this Sasquatch family have had contact with white people and learned to read and write in English although they had their own language as well. This may not be true of all Sasquatch people, but it is definitely true of this clan. Having read my book, or at least receiving it, and understanding who I am, through some inner sensing abilities, they named me, Water Woman.

Ocean learned much about the creatures upon her visit to Susan's home. Susan raised a large garden so that she could share some of her crops with the Sasquatch. She would prepare meals and place them on a picnic table in the woods for their enjoyment. In return, the Sasquatch would bring medicinal herbs to Susan and her family. They placed two medicine wheels on Susan's property: 'The Sasquatch said that these rock and shell and wood circles, describing the four directions, serve as Portals. You can go through them – but be aware! You may not be able to come back! The Sasquatch call this other world, the Sometime Place.'

Over time, Ocean came to know the Sasquatch family that frequented Susan's farm. There were 33 of them in all, some of whom were pregnant. Her encounters with them were filled with a sense of peace and love. She describes one such encounter:

Around me I begin to feel an aura of Love. It is a gentle frequency that expands a hundred feet on all sides. I recognize it as the same welcoming vibration I receive when the dolphin pods approach. Who is it? I wonder? I have the feeling someone is coming. There is something very loving in the air. Could it be the Wise Ones? This is the name for the Sasquatch people who are the most advanced spiritually and intellectually.

She also describes their appearance:

I ... become aware of two gentle dark eyes looking at me. They are not staring or piercing eyes, but soft, half-closed eyes. I feel the warmth behind them. Widening my gaze, I see the rest of the head, and a form standing just beneath the porch roof. In the fading light, I can see the head is big with curly hair surrounding the face. I hadn't pictured Sasquatch with curly hair. It looks very nice, like a child's favorite teddy bear, well-worn. I look steadily at the face to be sure I am seeing what I think is there. The head of this huge body is framed in the leaves from the trees? It seems to be merging with the leaves. I see a distinct curl over the right eye. There is no obvious neck, a dark outline of massive shoulders, long arms. The lower part of the body is too dark to see in detail, but it blocks out everything behind it, giving me an idea of its size. The eyes continue to stare softly at me. The head is below the height of the roof. I estimate this Sasquatch is more than six feet tall.

Sasquatch have many great abilities that would come as a surprise to most people, even to most Bigfoot believers. Ocean notes that they can walk 10 to 20 miles an hour. They can seemingly dematerialize and become invisible, whether by changing their own atomic structure or by creating an energy field that affects the ability of observers to see them, she does not know. It may also be that they can simply

> **❝ THEY CAN WALK AT 10 TO 20 MILES AN HOUR. THEY CAN SEEMINGLY DEMATERIALIZE AND BECOME INVISIBLE ❞**

cause humans to experience interrupted time, which would cause the humans to believe that a Sasquatch had dematerialized.

Sasquatch are also gifted in the use of what Ocean calls 'Infrasound,' this is sound which affects humans emotionally, mentally, physically, and multi-dimensionally. Infrasound can be used to communicate with others thousands of miles away. The sounds are of such low frequency that the human ear cannot detect them. Sasquatch may use them to create feelings of peace and calm or, when needed for protection, to produce feelings of terror and unease. Sasquatch also possess the ability to teleport objects by manifesting them and controlling them from a distance. This ability, and others, have developed among Sasquatch as they have evolved through time. Indeed, during the Ice Age, Ocean explains, the Sasquatch were much like us. They have subsequently evolved beyond us. They are our big brothers. One day we may be like them and may develop the great power that they possess.

Similar insights into the nature of Sasquatch are also offered by Ocean's friend and colleague, Jack 'Kewaunee' Lasperitis. Lasperitis claims that he is one of 76 individuals who have experienced the psychic Sasquatch and the extraterrestrial beings often connected with them. He claims to have had nearly 500 encounters in a 19-year period. The conclusion that he has reached based on these encounters is that Sasquatch are 'giant hairy *humans* with extraterrestrial origin' (3). Following years of looking for evidence of a giant prehistoric ape with limited intelligence, Lasperitis was shocked to discover that Sasquatch were much, much more.

He describes an encounter that began while he was sleeping and was gently awakened by a voice saying, 'Wake up, wake up, my friend; we are here.'

> Lying in bed facing the wall, I instantly opened my eyes and felt an *overwhelming* 'presence' in the bedroom. Turning around, I was fascinated beyond delight as my eyes beheld two living apparitions, very much 'alive,' of apelike men, which I immediately could see were the Sasquatch people. Without fear, I lay there and looked with great wonder, knowing I was privileged that these interdimensional nature-beings were sharing who they were by revealing their presence while in the astral state. Since September 1979, this non-hallucinatory experience has occurred to me a few hundred times. Sometimes it occurred in the company of multiple witnesses who experienced the same phenomenon, which, at times, included telepathic communication with these apparitions. (135)

Lasperitis experienced a deep sense of love and peace emanating from the two beings. The creatures were male. One was seven to eight feet tall, The other was around four feet. The smaller one seemed inexperienced, as if 'astral travel was new to him.'

> The larger creature had a remarkably interesting face, which looked half-ape, half-man! That face had an air of control and intelligence, and the eyes studied me in an inquisitive manner. In its effort to scrutinize me, I was surprised to observe the bottom half of the translucent body float through and below the bedroom floor until the being's face was only three feet from mine. This gave me a perfect view of the head. The aliveness and movement of the pupils in the eyes thoroughly intrigued me, indicating that I was indeed viewing something very extraordinary. (136)

Based on his experiences and the experiences of the other 75 percipients, Lasperitis believes that Sasquatch are human and not animals, that they are benign and not dangerous, that they have great psychic abilities, that they did not evolve on the earth, and that they are associated with UFOs and their occupants. Sasquatch can travel across dimensions at will. Sometimes they occupy a position between dimensions. In this state they can be seen by humans but do not have a physical presence:

> When a Sasquatch 'blips' into another dimension while walking on snow or in mud, the *proof* is that the physical tracks abruptly end as if the creature vanished into thin air – and, in reality, it truly did, by entering another, 'lighter,' less dense dimension. This phenomenon has been reported frequently by researchers and percipients. The sentient creatures use other dimensions both to hide from danger and to spy on humans while reading their minds. (161)

Lasperitis reports one such encounter with dematerializing Sasquatch. The report came from Oregon, where a snowmobiler was shocked to see three Sasquatch walking through the snow. The man became even more shocked when he realized that he could see through the Sasquatch – they were transparent. In addition, they were floating slightly above the snow, leaving no tracks.

Much of the work of the Sasquatch apparently takes place underground. Many people who have encountered Sasquatch have also reported strange, humming sounds coming from the ground. Lasperitis notes that this is because the Sasquatch live in underground cities. He relates the story of Mark, who was invited by a Sasquatch to visit one of these cities:

> Mark claims he was taken to a cave entrance that was very well camouflaged – in fact, he did not recognize it until they were almost beside the entrance. Another Sasquatch stood inside, guarding the entrance to the cave. Inside the cave Mark could see no visible lights, just a mysterious glow, which provided adequate light for them to see. (71)

To Mark's surprise, he was greeted by four tall Star People. These extraterrestrials appeared human, except for their tall stature. Two of the ETs were male, two were female. They spoke perfect English and were very polite. Mark was told that the Star People had been there for thousands of years and that their city ran for a mile and a half underground. They showed Mark a large amount of machinery operated by hairless beings, three to four feet tall. When Mark touched one of the tiny beings, he felt an electrical spark. He was convinced that they were clones, especially created to perform technical work. The Star People showed Mark how to view 'books' in their library, all of which were produced electronically upon a video screen. On another visit, Mark was escorted by Sasquatch to view a landed UFO.

Underground cities and flying saucers should not distract from the spiritual aspect of contact with the Sasquatch emphasized by both Lasperitis and Ocean, however – namely the spiritual enlightenment that their ancient wisdom can bring, from beyond the earth, from the deepest past, and from alternate dimensions. Joan Ocean describes her encounter with this aspect of the Sasquatch on her website:

> We settle into a deep meditation for contact, proceeding with the help of Medicine Woman, to enter the Portal. We experience the swirling energy, starting at our feet and slowly moving up, enveloping our bodies. At some point as each of us is ready, we leave the physical realm and travel with two Wise Ones through the Sometime Place into their alternate reality. There we meet with the Medicine Woman and she is happy to welcome us.
>
> I ask to see the future of Earthling and Sasquatch contact. I then experience a round room with many Ancient Ones, Sasquatch and Earth people who appear gentle and wise. We are learning many things from these Advanced Sasquatch people; how to live in splendor in their forests and 4th dimensional homes. They know about all the life forms in the woods. Every growing herb in the forest has a different nourishing and healing characteristic. We are even learning about herbs for Dream Walking, dematerializing, shape-shifting, spirit enhancing, voice and sound projecting, and entering other realities. They seem to recommend mind-enhancing herbs for 'out of body' experiences as a preparation for entering the 'invisible' worlds. If we can overcome our blocks to these inherent abilities, we can experience them. We have all the glands, consciousness and active DNA to do it.
>
> In this way we visit the inner earth people and other beings that live on planet Earth with us, although unseen by us. The Wise Ones know these places and these people.
>
> As we visit these places temporarily, I feel the sensation of being small, in the company of 7 foot Sasquatch and 9 foot Inner Earth people. They are gentle and loving, beaming their welcome. Their world is green, lush and peaceful. They are wisely discerning about who can enter in.
>
> This is just a small part of what I experienced. This Dream Walking is the kind of communication the Medicine Woman wants with Earthlings. We can enter their world with them. (www.joanocean. com)

*   *   *

It is not swirling, positive energy that I am now experiencing. This is not an alternate reality. This world is not green, lush and peaceful. I am not Dream Walking.

Smokey Crabtree opens the door to his garage and motions for us to enter. It is a cavernous building filled with all sorts of things, including a couple of boats and assorted lawnmower parts.

Along one wall are various animal mounts, most showing their age and covered with dust. I half expect to see a Bigfoot head protruding from the wall next to the stuffed beaver (normal sized, not giant).

> **❝ I'M NOT SURE WHAT THIS IS. IT'S NOT HUMAN, THAT'S FOR SURE ❞**

Smokey calls our attention to the far back wall of the building and throws open the plywood top on the plexi-glass box that houses his Fouke Monster carcass. There is a skeleton here all right, and it is real. There is stringy flesh attached to the bones – muscles and tendons in a state somewhere between preservation and decay, smelling of rot and formaldehyde.

Smokey props open the box with a plank and steps back so that we can get a better look.

The smell makes me want to gag, but I hold it back, not wanting to insult our host.

I move toward the box and take a closer look.

I'm not sure what this is. It is not human, that's for sure.

It is also not the three-toed biped of Fouke Monster lore.

My guess is that it is feline, some sort of very large cat. Surely it is too big to be an indigenous animal. No panther this large ever walked the woods of Arkansas.

Maybe it is a tiger – released into the woods for the hunting pleasure of some Texas oilman. That would explain the missing head. That would have been taken as a trophy.

But I'm not sure. How can I even think with that overpowering smell? There is one thing I do know, however. This is real. It is more real than the word of an eyewitness, more real than footprints in the Texas mud, dermal ridges or not. It is also clearly not a psychic phenomenon. It is earthy and material.

This is real. Real death. Real decay. Real stink.

But it is not Bigfoot.

*   *   *

11. Smokey's Monster, Fouke, Arkansas. Yeucch!

Asking if someone believes in Bigfoot is a lot like asking if they believe in Jesus. In both cases an affirmative answer tells you very little. In the case of Jesus, you have to follow up the question with several more: 'Are you Baptist, Catholic, Jehovah's Witness? Are you conservative or liberal?' With Bigfoot it is the same: 'Do you think that Bigfoot is an animal or a manimal? Is he a physical being or a spiritual being? Is he from the earth or from the stars?'

Some argue passionately that those who believe that Bigfoot is something more than a relic population of *Gigantopithecus* hurt the cause, making all Bigfoot believers look a little loony. Others argue that if Bigfoot science is going to put so much emphasis upon personal experiences and eyewitness accounts then it cannot ignore the incredible number of people who have claimed to experience Bigfoot as an intelligent, vocal, social being or as some sort of spirit being with powers far beyond those of humans. Sometimes the debate turns nasty. Each group accuses the others of being so blinded by their own preconceptions that they are unable to see the truth. One person rejects another's account because it does not match their own.

The Fouke Monster doesn't quite make it because it has three toes and because Smokey Crabtree claims to have its, obviously feline, carcass in a box. Jan Klement got a little too close to Kong for the comfort of some. He never should have told the story of how Kong had sex with a cow. Janice Carter Coy believes that the creatures can talk. Joan Ocean believes that they are spirit beings. For many in the Bigfoot community each of these versions of Bigfoot is suspect. The people making these claims are accused of being hoaxers or fools. The Bigfoot community is split into countless groups and organizations. It is Protestant sectarianism all over again.

When it comes to Bigfoot, just as with Jesus, there is a multitude of groups and individuals, each claiming to have possession of the one true faith.

The diversity of opinions about the nature of Bigfoot is probably a mark of the power of the belief in Bigfoot in the lives of these people. For a lot of people, Bigfoot is obviously more significant than lions and tigers and bears.

Bigfoot takes over lives.

Bigfoot moves in next door.

Bigfoot lives with families for generations.

Bigfoot speaks to the soul.

For some, Bigfoot is a forest god, leading our technological society away from our fallen ways and back toward the Garden of Eden. For others, Bigfoot is that secret part of ourselves, hidden from others, but central to who we are. For some, Bigfoot is the core of community, Bigfoot believers are like a church – people to gather with, who share our concerns, our hopes, and our dreams – people who share our fantasies and our realities.

And whatever the nature of Bigfoot turns out to be, whether physical or spiritual, whether terrestrial or cosmic, whether human or animal, whether fact or fantasy – the image of that hairy thing that lumbers out of the night is, for the true believers, clearly something beyond any of these dichotomies.

I don't know what that dead thing in Smokey Crabtree's box really is, but I do know that for many people, Bigfoot Lives!

PART TWO

Among the Arkansas Cave People

FOUR

Lost Worlds

The Bimini Islands lie just off the Atlantic coast of Florida, some 40 miles from Miami. This island paradise offers visitors the chance to fish, shop, boat, or just relax in the sun with a cool drink. Former residence of Ernest Hemingway, a man known for his ability to enjoy the good life, Bimini also claims to be 'the big game fishing capital of the world.'

As if that is not enough, promotional literature from the islands claims that it is believed that the island community was once part of the road system of the Lost Continent of Atlantis. It is the ideal place for me to pursue my research into the lost worlds of legend and, perhaps, to enjoy some much needed rest and relaxation.

I, however, am not in Bimini.

I am deep underground and chin-deep in muddy water that hovers at 58 degrees Fahrenheit. Except for the small glow of my helmet lamp, it is dark, very dark. A spotted salamander looks at me in disgust. I return the gesture. My boots are, once again, full of water. My jeans and shirt are soaked. I have just dropped down from a large cavern into a tiny compartment, roughly the size of a shallow grave. My heart is racing and I am beginning to feel a little panicked. I hear the voices of my companions from above me, to my left and my right, but I cannot see them. It sounds as if I am listening in on a conversation in another room, or in another reality. Perhaps this is the way conversations sound to the deceased, just before the dirt is thrown on the grave. 'I heard a fly buzz when I died,' and all of that. I feel completely alone. The salamander blinks and I take a deep breath and begin sliding cautiously to my right. I have been told that I have to move about ten feet before I will see the opening, slightly above my head, that leads to another cavern and to the company of my friends. The ten feet seem like an eternity. I take one step at a time. Now I can see a light shining from above, the glow of friendly flashlights. For a moment I hesitate. The advice that I have always heard for *handling this type of situation comes quickly to mind: 'Don't go into the light!'*

I take another breath and press on.

My wife has never understood why I went searching for Bigfoot in the swamps of Texas and in a smelly Arkansas garage instead of in the majestic forests of the Pacific Northwest. She only looked at me with a bemused smile when I told her that my search for lost civilizations was taking me, not to some island paradise, but to a cave in Cushman, Arkansas and deep into the bowels of the earth.

PLATO'S ATLANTIS

Plato's *Timaeus* tells the legend of Atlantis, a great and powerful island civilization whose navies sought to conquer the peoples of the Mediterranean. According to Plato, this civilization met a terrible end as a result of a series of natural disasters that caused it to sink into the sea. In *Critias*, Plato expounds upon the legend of the lost continent and recounts tales of a great war between Athens and Atlantis, which preceded its destruction. According to Plato, the Atlantean civilization enjoyed great riches, both those found within its own fertile boundaries, and those received by trade with other peoples. The capital city of Atlantis was surrounded by alternating concentric circles of land and water. Great bridges and canals were built from

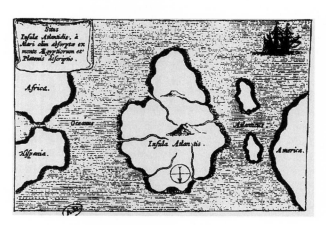

12. A map of Atlantis. Notice that north is down and south up, a reversal of the modern standard.

the center island to the outermost rings, making transportation easy between the great metropolis, with its royal palace and temple of Poseidon, and the lands beyond. With fountains flowing with hot and cold water, luxurious dwellings, and beautiful weather, Atlantis is described by Plato as an ancient paradise – a paradise fitted with a mighty navy and driven by great ambitions. Alas, the civilization was not to last. Its great strength and beauty were lost to the forces of earth and sea. It sank beneath the waves, lost to history and never to return.

Well, maybe not.

Atlantis, it seems, has enjoyed quite a bit of attention, despite its disastrous end and its failure to find a proper place in respectable versions of history. While much of that attention has come from those who treat Plato's story as a myth or an allegory, like Francis Bacon in his *New Atlantis*, it has also received a great deal of attention as something much more.

THE ANTEDILUVIAN WORLD

Ignatius Donnelly's *Atlantis: The Antediluvian World* was published in 1882 by the former Congressmen from the State of Minnesota. Apparently having taken advantage of the availability of the Library of Congress while he was in Washington, D.C., Donnelly published his influential book on Atlantis shortly after his retirement from public life. Though there had been books on Atlantis before Donnelly's, none managed to gain the popular attention that his did, especially in the English-speaking world. Other modern speculators had identified Plato's Atlantis with various ancient cultures, perhaps most importantly with the ancient cultures of the Americas, so Donnelly was certainly not the first in that regard. His work, however, had the benefit of being well written and accessible to the average reader of the day.

Donnelly's approach to his subject is quite bold. He begins his book with a statement of claims that he intends to demonstrate. It is quite an ambitious list. There are 13 claims in all:

1. That there once existed in the Atlantic Ocean, opposite the mouth of the Mediterranean Sea, a large island, which was the remnant of an Atlantic continent, and known to the ancient world as Atlantis.

2. That the description of this island given by Plato is not, as has been long supposed, fable, but veritable history.

3. That Atlantis was the region where man first rose from a state of barbarism to civilization.

4. That it became, in the course of ages, a populous and mighty nation, from whose overflowings the shores of the Gulf of Mexico, the Mississippi River, the Amazon, the Pacific coast of South America, the Mediterranean, the west coast of Europe and Africa, the Baltic, the Black Sea, and the Caspian were populated by civilized nations.

5. That it was the true Antediluvian world; the Garden of Eden; the Gardens of the Hesperides; the Elysian Fields; the Gardens of Alcinous; the Mesomphalos; the Olympos; the Asgard of the traditions of the ancient nations; representing a universal memory of a great land, where early mankind dwelt for ages in peace and happiness.

6. That the gods and goddesses of the ancient Greeks, the Phœnicians, the Hindoos, and the Scandinavians were simply the kings, queens, and heroes of Atlantis; and the acts attributed to them in mythology are a confused recollection of real historical events.

7. That the mythology of Egypt and Peru represented the original religion of Atlantis, which was sun-worship.

8. That the oldest colony formed by the Atlanteans was probably in Egypt, whose civilization was a reproduction of that of the Atlantic island.

9. That the implements of the 'Bronze Age' of Europe were derived from Atlantis. The Atlanteans were also the first manufacturers of iron.

10. That the Phœnician alphabet, parent of all the European alphabets, was derived from an Atlantis alphabet, which was also conveyed from Atlantis to the Mayas of Central America.

11. That Atlantis was the original seat of the Aryan or Indo-European family of nations, as well as of the Semitic peoples, and possibly also of the Turanian races.

12. That Atlantis perished in a terrible convulsion of nature, in which the whole island sunk into the ocean, with nearly all its inhabitants.

13. That a few persons escaped in ships and on rafts, and, carried to the nations east and west the tidings of the appalling catastrophe, which has survived to our own time in the Flood and Deluge legends of the different nations of the old and new worlds. (1–2)

Whew ... Donnelly not only seeks to show that Plato's story of Atlantis describes a historical island nation but also claims that Atlantis represents the source of human civilization: its technology, religion, culture, and art.

Donnelly's approach is multifaceted. He argues that Plato's story of Atlantis was intended as an historical account. Then he turns his attention to the question of whether or not such a disaster, as described by Plato, was possible. He examines evidence for the lost continent on the floor of the Atlantic Ocean. He compares the flora and fauna of the new world with that of the old to show how a mid-Atlantic continent could have been the source of the cross-propagation of species. Donnelly explores the similarities among flood accounts from various cultures and suggests that the story of the fall of Atlantis is at their source. He compares the religion and culture of the old and new worlds, finding in their similarities evidence of a common origin.

❝ ATLANTIS PERISHED IN A TERRIBLE CONVULSION OF NATURE ... THE WHOLE ISLAND SUNK INTO THE OCEAN ❞

Finally, Donnelly describes his reconstruction of the civilization of the Atlanteans. The Atlanteans, he tells the reader, were residents of a great island nation, with smaller islands forming 'stepping stones' to Europe and Africa to the east and the Americas in the west. The central island was marked by volcanic peaks, rising to 1,500 feet. There were four rivers on the island, flowing north, south, east, and west from the center. The people themselves were of two different races. One was a race of dark complexion whose descendants spread to Central America and Egypt. The other was a fair-skinned race, the progenitors of the people of northern Europe. They worshipped an omnipotent creator god and saw the sun as the emblem of the deity. The religion of the Atlanteans had an ordered priesthood. The civilization of Atlantis was prosperous and strong, ruled by a monarchy and structured by law. Their trade ships and their navies crossed the Atlantic, east and west, and sailed the waters of the Mediterranean. Their culture was truly the highest of the ancient world:

They had processions, banners, and triumphal arches for their kings and heroes; they built pyramids, temples, round-towers, and obelisks; they practised religious ablutions; they knew the use of the magnet and of gunpowder. In short, they were in the enjoyment of a civilization nearly as high as our own, lacking only the printing-press, and those inventions in which steam, electricity, and magnetism are used. (323)

The loss of the great empire must have sent shock waves around the world. Atlantis passed, but not before its influence had transformed the peoples and cultures of the earth.

Donnelly's understanding of Atlantis is a product of his quest for an answer to the mysteries of cultural origins. Seeing similarities in the religion, technology, and art of cultures separated by vast oceans, he seeks a common source that would explain these similarities. For him, Atlantis seems the most logical explanation. Attested to by Plato, Atlantis is compelling to Donnelly because its existence can provide an explanation of why cultures in Egypt and the Americas each built magnificent pyramids, for example. Donnelly's knowledge of Atlantean culture is thus arrived at through a bit of reverse engineering. The culture of Atlantis, by definition, must have possessed those qualities that seem to be a part of the common culture of humanity. Since religion is a cultural universal, Atlantis is the source of all religion, its Atlantean form representing the source from which other religions grew. Donnelly took his argument, and it was taken by many of his readers, to be a rational and scientific theory of cultural origins. Donnelly's theory was weird but, to Donnelly and many of his readers, it was also science.

MU, LEMURIA AND ATLANTIS

At the same time that Donnelly was propagating his ideas of Atlantis, Augustus Le Plongeon was offering another version of the story, with the island nation of Atlantis taking the name of Mu (pronounced Moo). Le Plongeon believed that his translation of ancient Mayan texts revealed a new-world version of the Atlantis story, giving more credence to Plato's old-world version. Like Plato's Atlantis, Le Plongeon's island of Mu disappeared beneath the surface of the Atlantic Ocean in the distant past. In the early twentieth century, James Churchward argued that the story of Mu was actually the story of a second lost-island civilization. Atlantis had disappeared into the Atlantic Ocean, Mu into the Pacific. This lost island of the Pacific also came to be known as Lemuria.

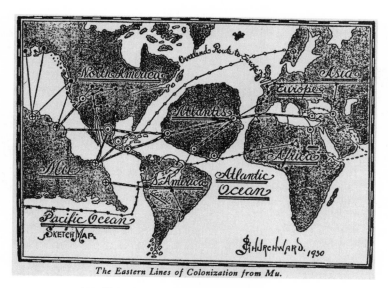

The Eastern Lines of Colonization from Mu.

13. Churchward's map of Atlantis and Mu.

The name 'Lemuria' was coined at the end of the nineteenth century by geologists and biologists as an attempt to explain the fact that lemurs are not only found on the island of Madagascar but also on the Indian subcontinent and on the islands of Malaysia. A prehistoric land bridge uniting these now separate lands would offer a neat explanation of how the lemur population managed to get from one location to the other – a feat that looks impossible if it is assumed they had to cross great bodies of water. The theoretical land bridge was therefore called 'Lemuria.' Its existence was deduced in a fashion similar to that used by Donnelly to deduce the existence of Atlantis – with lemurs playing the role of cultures.

The most developed vision of lost Mu/Lemuria is found in the writings of theosophist Helena P. Blavatsky and her followers. According to Blavatsky, the true story of Atlantis and Lemuria can be found in an ancient manuscript known as *The Book of Dyzan*. This book was revealed to Blavatsky by Tibetan masters – teachers and practitioners of the ancient arts of humanity. Blavatsky's commentaries and extrapolations on *The Book of Dyzan* were published in 1888 as *The Secret Doctrine*, one of the classics of theosophy. According to *The Secret Doctrine*, life evolved on earth through a procession of stages.

In each stage of development, humanity took on different forms and characteristics. Each instantiation of humanity is known as a 'Root Race.' The history of humanity is the story of the progress of our species through seven stages or seven Root Races. We live today at the fifth stage. The sixth and seventh are yet to come. Each Root Race is associated with a different continental homeland.

Human history began in 'The Imperishable Sacred Land.' The Imperishable Sacred Land has never suffered the fate of other continents that rise and pass away. This land will survive from the beginning of human history to the end. Blavatsky writes in *The Secret Doctrine*: 'Of this mysterious and sacred land very little can be said, except, perhaps, according to a poetical expression in one of the Commentaries, that the "pole-star has its watchful eye upon it, from the dawn to the close of the twilight of 'a day' of the GREAT BREATH"' (6). According to some readers of Blavatsky's difficult prose, the Imperishable Sacred Land is located at the North Pole, a proposition that might have seemed more likely in the 1880s than it does today. Other readers have suggested that this far-northern land is actually in the interior of the earth, reached by an opening at the pole (but more about this in the next chapter). According to Blavatsky's follower, W. Scott-Elliot, in his *The Lost Lemuria* (1904), the physical bodies of the First Root Race of the Imperishable Sacred Land would have looked to modern humans like giant phantoms, assuming that we could see them at all, their bodies consisting of 'astral matter.'

The second continent is identified by Blavatsky as 'Hyperborea.' Hyperborea was a land that 'stretched out its promontories southward and westward from the North Pole to receive the Second Race, and comprised the whole of what is now known as Northern Asia' (7). Winter was unknown to this northern continent in the ancient days, perhaps because the earth had not yet tilted on its axis. According to Scott-Elliot, Hyperboreans were a bit more substantial than their ancestors, but still essentially shapeless. They had rudimentary organs and skeletal systems and reproduced through asexual budding. Like the First Root Race, the second would have also been invisible to the human eye. Hyperborea eventually broke apart, the lands of the Arctic Circle being its last remains.

The third continent Blavatsky called 'Lemuria.' Lemuria was a land mass that extended from the Indian Ocean to Australia. Scott-Elliot notes that, though closer to humans than the Hyperboreans, the Lemurian race should best be 'regarded rather as an animal destined to reach humanity than as a human according to our understanding of the term' (20). During the Lemurian age, the Third Root Race interbred with the most developed members of the animal kingdom, huge ape-like creatures that existed on Mars, earth, and Mercury:

> From the Etheric Second Race, then, was evolved the Third – the Lemurian. Their bodies had become material, being composed of the gases, liquids and solids which constitute the three lowest sub-divisions of the physical plane, but the gases and liquids still predominated, for as yet their vertebrate structure had not solidified into bones such as ours, and they could not, therefore, stand erect. Their bones in fact were pliable as the bones of young infants now are. It was not until the middle of the Lemurian period that man developed a solid bony structure. (22)

Over time, the Lemurians became more substantial and more like modern humans. Scott-Elliot describes a typical Lemurian at a later point in their development:

> His stature was gigantic, somewhere between twelve and fifteen feet. His skin was very dark, being of a yellowish brown colour. He had a long lower jaw, a strangely flattened face, eyes small but piercing and set curiously far apart, so that he could see sideways as well as in front, while the eye at the back of the head – on which part of the head no hair, of course, grew – enabled him to see in that direction also. He had no forehead, but there seemed to be a roll of flesh where it should have been. The head sloped backwards and upwards in a rather curious way. The arms and legs (especially the former) were longer in proportion than ours, and could not be perfectly straightened either at elbows or knees; the hands and feet were enormous, and the heels projected backwards in an ungainly way. The figure was draped in a loose robe of skin, something like rhinoceros hide, but more scaly, probably the skin of some animal of which we now know only through its fossil remains. Round his head, on which the hair was quite short, was twisted another piece of skin to which were attached tassels of bright red, blue and other colours. In his left hand he held a sharpened staff, which was doubtless used for defence or attack. It was about the height of his own body, viz., twelve to fifteen feet.

❬ IN HIS RIGHT HAND . . . HE LED A HUGE AND HIDEOUS REPTILE, SOMEWHAT RESEMBLING THE PLESIOSAURUS ❭

In his right hand was twisted the end of a long rope made of some sort of creeping plant, by which he led a huge and hideous reptile, somewhat resembling the Plesiosaurus. The Lemurians actually domesticated these creatures, and trained them to employ their strength in hunting other animals. The appearance of the man gave an unpleasant sensation, but he was not entirely uncivilised, being an average common-place specimen of his day. (24)

Scott-Elliot notes that the pineal gland of modern humans is an atrophied remnant of the third eye which itself may be employed for a kind of astral vision. Lemuria was destroyed by volcanic activities.

The fourth continent was the continent of Atlantis. The land of Atlantis was inhabited by a Root Race fully human in appearance. Blavatsky indicates that Atlantis should be regarded as the first historical continent, except that most historians ignore the ancient record of its existence. Scott-Elliot offers detailed descriptions of Atlantean life and culture in *The Story of Atlantis* (1896). For the first time, humanity exhibited cultural development, including education, art, science, and religion. Scott-Elliot describes the education offered to the exceptionally gifted children of Atlantis as including training in the use of psychic abilities and the occult properties of plants, metals, and precious stones to heal diseases. They were also taught the alchemical techniques of matter transmutation as well as the ability to tap into the occult powers of the universe. He describes the marvelous technological developments that were achieved by the Atlanteans, including air-ships and flying machines.

These air-ships were for the benefit of the wealthier class. They were generally built for a small number of persons, from two-seaters to ships with room for eight. As the Atlantean age descended into warfare, these ships were used for battle. These battle-ships were much larger, accommodating up to 100 sailors. Scott-Elliot describes their construction:

The material of which the air-boats were constructed was either wood or metal. The earlier ones were built of wood – the boards used being exceedingly thin, but the injection of some substance which did not add materially to the weight, while it gave leather-like toughness, provided the necessary combination of lightness and strength. When metal was used it was generally an alloy – two white-coloured metals and one red one entering into its composition. The resultant was white-coloured, like aluminium, and even lighter in weight. Over the rough framework of the air-boat was extended a large sheet of this metal, which was then beaten into shape, and electrically welded where necessary. But whether built of metal or wood their outside surface was apparently seamless and perfectly smooth, and they shone in the dark as if coated with luminous paint. (68)

The craft could travel at an elevation of several hundred feet and approached speeds of 100 miles an hour.

Unlike the earlier Lemurian Root Race, the Atlantean culture was advanced enough to have included religion. The Atlanteans believed in a Supreme Being, symbolized by the sun. The worship of this sun-deity took place on hill tops where circles of upright monoliths were built. These monoliths, of which Stonehenge is an extant example, were also used for astronomical observances. Atlantis, however, suffered a period of cultural debasement before it was finally destroyed. Peace and prosperity gave way to war and violence, and the religion of the sun at times descended into fetishism. Atlantis was finally destroyed when it sank into the Atlantic Ocean.

The case of Stonehenge provides some interesting insight into the fall of the Atlantean Empire. According to Scott-Elliot, Atlanteans first arrived in Europe 100,000 years ago, arriving first on the shores of Scandinavia. These early Atlantean settlers were of the Akkadian race, taller and fairer than the indigenous Europeans. Scott-Elliot's description of these settlers bears some resemblance to the Puritan settlers in North America who left Europe as a form of religious protest to establish a more simplistic and austere religious community. He writes: 'The rude simplicity of Stonehenge was intended as a protest against the extravagant ornament and over-decoration of the existing temples in Atlantis, where the debased worship of their own images was being carried on by the inhabitants.' (53)

This theosophical version of Atlantis was decidedly weirder than Donnelly's version. But, we should note, it also purported to be a scientific account. Blavatsky and her followers not only claimed that their knowledge was based upon ancient manuscripts, itself a claim to empirical validation of sorts, but also appealed to the developing fields of evolutionary biology and geology. Specifically, Blavatsky and other theosophists were greatly influenced by the evolutionary monism of German biologist and champion of Charles Darwin, Ernst Haeckel. Haeckel's famous claim that 'ontogeny recapitulates phylogeny' can be seen as a critical source for theosophical theories of human evolution, with the earlier Root Races looking like gigantic zygotes and the Lemurians looking like giant, late-term fetuses dressed in rhinoceros hide. When your science is weird enough, no beliefs are too bizarre.

Mount Shasta

In addition to Scott-Elliot, Frederick S. Oliver also contributed to the developing theosophical picture of life in ancient Atlantis. Oliver claimed to have channeled the text of the manuscript for his book *A Dweller on Two Planets* through a process of automatic writing. *A Dweller on Two Planets* was published posthumously in 1905 by Oliver's mother. The book purports to be the story of Phylos the Thibetan. According to Oliver's preface, what he calls the 'Amanuensis' Preface,' Oliver received his revelations in the proximity of California's Mount Shasta. There he encountered the psychic presence of Phylos, mostly through spoken voice alone, but sometimes also by sight. *A Dweller on Two Planets* tells the story of Phylos through multiple incarnations, including a life in the ancient land of Atlantis and upon the planet Venus. According to Phylos,

> The glories and marvels of Atlantis the Great were not in vain. Thou and I, reader, lived then, and before then. The glories of those long-dead centuries seen by us have lived enshrined in our souls, and made us much, aye, most, of what we are, influenced our acts, soothed us with their beauty. (87)

Like Blavatsky and Scott-Elliot, Oliver/Phylos asserts that the ancient Atlanteans were sun-worshippers. Worship of the sun should not be taken as some sort of fetishism, however, for the Atlanteans knew that the sun only symbolized the one true god, known as Incal.

> Truly, I have said that we believed in Incal, and symbolized him as the Sun-God. But the sun itself was an emblem. To assert that we, despite our enlightenment, adored the orb of day, would be as absurd as to say that the Christians adore the cross of the crucifixion for itself; in both cases it is the attached significance that caused the sun, and causes the cross, to be held in any sort of regard. (88)

Oliver/Phylos also describes an Atlantean air-ship, which he calls a 'vailx.' According to his account, the craft came in four standard lengths: 25 feet, 80 feet, 155 feet, and 355 feet. The vessels were cigar-shaped, tapering to sharp points at either end. Usually there was an open promenade deck at one end of the craft and portholes of crystal in rows along the side, top and floor. He describes a journey aboard one of these craft, noting that the passengers would sit on the open promenade deck, dressed in warm arctic attire, and enjoy the view, some two miles above the surface of the earth. Passengers aboard the great vailxa could communicate with friends and family at great distances through the use of viewing screens. The ships were furnished with libraries, musical instruments, potted plants, and small birds.

Oliver/Phylos also notes that the craft were capable of leaving the air and traveling underwater. A transition from air to water is described in terms reminiscent of 'the flying sub' in Irwin Allen's classic television series *Voyage to the Bottom of the Sea*:

> For change we decided to forsake the realms of the air for those of the deep where the shark is king. Like all vailx of the class to which it belonged, ours was constructed for both aerial and submarine service, the plates of the sliding deck and the other movable parts of the hull being capable of very close approximation by means of setscrews and rubber washers. To settle straight down into the ocean would be too much like a landing on terra firma. But being at a height of two miles, more or less, the conductor was directed to gradually reduce the repulsion current, thus diminishing our buoyancy so as to bring us into the water ten miles distant from where the slant commenced. He was further ordered to do this while maintaining a speed which would, though very slow for a vailx, be really swift, that is, he was to cover ten miles in as many minutes. When we struck the water at this rate of progress the shock which the entering needle experienced was sufficiently great to cause its inmates to stagger, and little exclamations were made by the ladies. (175–6)

Despite the intriguing details offered concerning life in Atlantis, readers of *A Dweller on Two Planets* seemed most interested in Oliver's descriptions of California's Mount Shasta. As an interlude between parts one and two of his book, Oliver includes what he calls 'Seven Shasta Scenes.' These mysterious scenes would prove to be quite influential. In the seventh of these he writes:

> What secrets perchance are about us? We do not know as we lie there, our bodies resting, our souls filled with peace, nor do we know until many years are passed out through the back door of time that that tall basalt cliff conceals a doorway. We do not suspect this, nor that a long tunnel stretches away, far into the interior of

majestic Shasta. Wholly unthought is it that there lie at the tunnel's far end vast apartments, the home of a mystic brotherhood, whose occult arts hollowed that tunnel and mysterious dwelling: 'Sach' the name is. Are you incredulous as to these things? Go there, or suffer yourself to be taken as I was, once! See, as I saw, not with the vision of flesh, the walls, polished as by jewelers, though excavated as by giants; floors carpeted with long, fleecy gray fabric that looked like fur, but was a mineral product; ledges intersected by the builders, and in their wonderful polish exhibiting veinings of gold, of silver, of green copper ores, and maculations of precious stones. Verily, a mystic temple, made afar from the madding crowd. (248)

In this passage Oliver hints at, rather than describes, a dwelling deep inside Mount Shasta. He calls it 'the home of a mystic brotherhood.' The walls are like polished jewels, the floors are carpeted with a fur-like mineral. It is a mystic temple.

Apparently, this enigmatic passage was the inspiration for the Rosicrucian William Spencer Lewis' 1925 article 'Descendants of Lemuria: A Description of an Ancient Cult in California,' published in *The Mystic Triangle* under the pseudonym 'Selvius.' In this article, Lewis claims that Mount Shasta is the home of the last descendants of the ancient Lemurians, whose village is 'nestled at the foot of a partially extinct volcano.' The village is secluded and protected by an invisible boundary so that only four or five strangers have ever set foot there. Lewis claims that the number of strange experiences reported by visitors to the area is sure evidence of the presence of some mystical force. In particular, he claims that Professor Edgar Lucin Larkin, director of the Mount Lowe Observatory, had seen the temple of the mystic village while examining the area through his telescope. He also notes that at one time a delegate from the mystic village had visited San Francisco and that tall, white-robed, gray-haired, barefoot saints had been seen

❝ A DELEGATE FROM THE MYSTIC VILLAGE HAD VISITED SAN FRANCISCO . . . TALL, WHITE-ROBED, GRAY-HAIRED BAREFOOT SAINTS HAD BEEN SEEN ❞

on the highways and in the streets throughout the area, occasionally even shopping in local stores, paying for their purchases with gold nuggets. Eyewitnesses have also reported strange boats that sail upon the Pacific Ocean only to fly through the air to the mountain. Lewis' theory was expounded on in his book *Lemuria: The Lost Continent of the Pacific* in 1931, written under the name Wishar Spenle Cerve.

Another theosophist influenced by Oliver, and who would contribute to the Mount Shasta mythos, was Guy Warren Ballard, writing under the name of Godfre Ray King. Ballard's *Unveiled Mysteries* was published in 1934. In this book, Ballard describes his meetings with Saint Germaine, an ascended master, on the slopes of Mount Shasta, and their subsequent journeys into past lives in which they both lived in Mu and Atlantis. Like *A Dweller on Two Planets*, King describes Venusians as advanced masters and reports that he and Saint Germaine encountered Venusian visitors to earth. Ballard's books proved to be quite influential and led to the beginning of the religious movement known as 'The Ascended Masters' I AM Activity,' which today claims over 300 local groups around the world and was very influential in the teachings of new-age guru Elizabeth Clare Prophet.

Clearly the most mainstream example of the influence of the legend of Mount Shasta's Lemurian brotherhood is found in James Hilton's *Lost Horizons*. In this novel Hilton describes a mysterious and hidden Tibetan monastery. Though clearly based on the Tibetan Buddhist 'Shambala,' a mystical kingdom hidden in the heights of the Himalayan mountains, Hilton's Shangri La is also related to Oliver's and Lewis' Mount Shasta Lemurians.

John Flinn noted in the *San Francisco Chronicle* (August 1, 2004) that, when asked by a reporter in 1941 to name the real place most like the fictional Shangri La, Hilton named Weaverville, California. Weaverville is located at the base of Mount Shasta, a gold rush town of the mid-nineteenth century. According to Flinn, Hilton's claim is not as strange as it sounds because Weaverville is the home of the Taoist Joss House, the oldest Chinese temple in the state of California, built in the 1850s. Of course, what Flinn doesn't mention are the occult legends of a Lemurian village in that very area.

A man calling himself 'Peter Mt. Shasta' has seized upon this connection to make explicit claims about the identity of Shangri La and Shambala with Mount Shasta. Peter Mt. Shasta writes:

> What is Shambhala? From the beginning of time legends have come down to us of a semi-mythical place in Asia inhabited by a race of highly evolved beings living in peace and harmony and working for the benefit of humanity. *Lost Horizons*, and the classic movie by the same title, portrayed such an idyllic place called Shangri-La. There, in

14. Ronald Colman relaxes in Shangri La. Was he in Tibet or Northern
California? In any event, 'It's astonishing and incredible,
but . . . you're still alive, Father Perrault!'
From Frank Capra's movie of *Lost Horizon* (1937).
Image courtesy of the Kobal Collection.

a lost world beyond the Himalayas, lived a race of enlightened beings who had so perfected themselves that they ceased to age and lived in youthful, healthy bodies, enjoying the beauty of life to the fullest, radiating love to the rest of the world. (www.mountshastamagazine.com)

Mt. Shasta goes on to claim that it was when visiting the Mount Shasta area that Hilton's vision of Shangri La took shape. Perhaps, he suggests, Hilton perceived something from another time or another dimension when he visited there. Mt. Shasta also notes that Hilton was not the first to experience mystical energies on the mountain. Some have seen white-columned temples resembling the Parthenon on the slopes of the mountain, or been visited while camping on the mountain by tall beings in long robes who spoke of an ancient race who lived on the far side of Mount Shasta, from where they journeyed out into the world on missions of guidance, peace, and healing.

Mt. Shasta describes his own experience on the mountain:

Several years ago while meditating one morning, Trungpa Rinpoche appeared to me and took me into the blue sky above Mount Shasta. He pointed toward the Shasta Valley, and with a sweep of his arm that encompassed the area from Mount Shasta to Mount Ashland, said, 'This is the New Shambhala.'

I saw that many people would be coming to the area, drawn by a similar vision. I saw many spiritual centers, temples and retreats, domes, new types of buildings constructed in harmony with nature to house this inflow of seekers that the Masters would invite. And I felt elated to be alive at the time of this Great Awakening, to be a part of the building of the New Shambhala.

*   *   *

It sounds lovely, this Californian Shangri La, nestled at the foot of a great mountain. The Weaverville, California website describes vast tracts of forested land, and claims that 'tucked away among fir and pine forested slopes, wildflowers in season cloak remote dells with splashes of brilliant color. Clear, tumbling waters from tarns and snow packs high up under towering peaks course down through rock walled canyons.' It truly sounds like Shambala, like the land of the ancient Lemurians, like the Garden of Eden. A worthy location to search for evidence of lost worlds.

Or so it seems to me as I am lying on my back on a floor of wet, cold, slippery mud. The ceiling to the cave is too low for walking or crawling and the mud is too slippery to slide along on my belly. When I tried that I got nowhere. My back to the muddy floor, I lift my feet to the rock ceiling to try and propel myself along. I make no progress. Instead I simply spin in place, counter-clockwise in the mud. In a moment of clarity I see this as a metaphor for the investigation that led me to this cave, an investigation that kept going around in circles and which finally led me into the subterranean depths of lost world theories.

I was first drawn to this location when I encountered an unusual story tucked away in the corners of two or three obscure websites. The story told of several encounters between explorers and members of an unknown but advanced civilization. At least two types of beings had been encountered in or around a cave in northern Arkansas. The first was described as standing seven to eight feet tall with skin of pale blue. The second was a Bigfoot-like creature known to chase explorers from the area by throwing rocks. Having had some experience with rock-throwing Sasquatch, I, of course, had to investigate.

Finding information about the eyewitnesses and the location proved to be quite difficult, however, though I was finally led to an internet chat group, known as the Shaver Mystery Group, whose members seemed to have some information about the story. I never should have told them that I was researching a book, however, especially not one titled Weird Science and Bizarre Beliefs. *When I suggested that the title was meant as a tribute to the pulp magazines in which many of the alternative theories I am interested in were discussed, they seemed a little friendlier. Then one of the group challenged the pedigree of the word 'bizarre' as used in reference to esoteric phenomena. Could I name an instance when the term 'bizarre' had been used as a title for one of the magazines in question, he asked.*

I first mentioned the current British publication that goes by the name Bizarre, *but this proved unacceptable. Too current, too glossy, too ironic.*

What about the fetish magazine from the 1940s and 1950s? Pulp enough, he replied, but rather tasteless. It was an 'under-the-counter publication' and a lot closer to pornography than true esoterica.

Bizarre Mystery, Bizarre Fantasy Tales, Bizarre Bazaar, Bizarre Sex and Other Crimes of Passion? *These magazines mostly published fiction.*

Okay. How about this one? Bizarre. *It was published once in 1941 and included a story by H.P. Lovecraft. I know it is a sci-fi magazine, but it is very, very, very obscure and very, very, very pulpy.*

That was the key. My interlocutor sent me an image of the cover of the magazine in question. I had passed the test. They would tell me what I wanted to know. I could now continue my quest for the lost world.

FINDING THE LOST ATLANTIS

On October 20, 1912 the discovery of Atlantis was announced in the pages of the *New York American*. Paul Schliemann, whose grandfather Heinrich Schliemann had excavated the remains of ancient Troy, announced that his grandfather had left information in regards to the location of lost Atlantis. The junior Schliemann claimed that after years of investigation and research he was prepared to announce his findings publicly. His account was entitled:

How I Found the Lost Atlantis, The Source of All Civilization

In The Most Astonishing Scientific Narrative Ever Published the Grandson of Troy's Discoverer Tells Why He Believes He Has Unravelled the Greatest World Mystery

by Dr. Paul Schliemann

Grandson of Dr. Heinrich Schliemann, Who Discovered and Excavated Ancient Troy and Other Great Cities of the Mycenean Civilization, Which Preceded and Was Greater Than That of the Greeks

According to Schliemann, Atlantis is identified with the Dolphin Ridge, an underwater plateau, between 25 and 50 degrees north, and 20 to 50 degrees west. The Azore Islands represent the mountains of Atlantis, the only part of that great land that now rises out of the waves. Schliemann was led to the location by his grandfather's posthumous instructions. The elder Schliemann was led to Atlantis by a chance discovery during his excavation of Troy. Heinrich wrote in his instructions to his grandson:

> When in 1873, I made the excavation of the ruins of Troy at Hissarlik and discovered in the Second City the famous 'Treasure of Priam,' I found among that treasure a peculiar bronze vase, of great size. Within it were several pieces of pottery, various small images of peculiar metal, coins of the same metal and objects made of fossilized bone. Some of these objects and the bronze vase were engraved with a sentence in Phoenician hieroglyphics. The sentence read 'From the King Chronos of Atlantis.' (http://www.sacred-texts.com/atl/hif/index.htm)

Later, Heinrich discovered matching pottery at the Louvre, this time from an excavation in Central America. Matching vases discovered in the Mediterranean and in Central America pointed toward a common origin in some central location – Atlantis!

Carefully following the leads left by his grandfather, Paul followed the trail to Egypt and South America, learning to read hieroglyphics and deciphering the clues until his conclusion was foolproof. The next step was simply to explore the undersea city and uncover its riches. The newspaper included an illustration of exotic diving equipment labeled: 'The New Diving Armor, Designed by Chevalier Pino to Resist Enormous Water Pressure, Which Will Be Used by the English Expedition Which Has Set Out to Find the Treasures of Atlantis' (http://www.sacred-texts.com/atl/hif/index.htm).

Alas, the results of Schliemann's expedition were never reported, neither in the *New York American* nor any other newspaper or journal. One suspects that the newspaper, like many others operated by William Randolph Hearst, intended the story to be little more than a joke.

Schliemann was not the only person to claim knowledge of the site of lost Atlantis, however, nor was he the most influential person to do so. That distinction belongs to the 'Sleeping Prophet,' Edgar Cayce. Cayce rose to prominence in the early part of the twentieth

15. Dr. Paul Schliemann's diving armor,
designed, according to the *New York American*, to resist enormous water
pressure: mandatory in searching for lost Atlantis.

century, receiving great attention following an October 9, 1910 article reporting his mystic abilities, published in the *New York Times* concerning a presentation of his healing abilities before a Boston research society. According to the claims, Cayce could tune into a person's physical ailments, making diagnoses and suggesting cures, with only the patient's name and address at his disposal – even if the patient were separated from him by a great distance.

In addition to his work as a psychic diagnostician, Cayce also performed 'life readings' in which he would enter into a sleeping trance and gather information concerning an individual's past lives. Many of these life readings concerned past lives lived in Atlantis. Following his death in 1945, several thousand pages of stenographic recordings of these life readings were discovered. They covered a 40-year period and included readings for 8,000 people.

According to students of Cayce's readings, many people living in America in the twentieth century were reincarnated Atlanteans. According to Cayce's readings, Atlantis was a technologically advanced civilization. Indeed, modern technology is, in many ways, simply a rediscovery of ancient Atlantean technology. Seeing it as a warning for present-day civilization, Cayce noted that many Atlanteans fell under the sway of technology and power, losing their souls in the bargain. Becoming so attached to the world of materiality, they lost touch with the world of the spirit. The Atlanteans who followed the ways of material culture, the Sons of Belial, soon came into conflict with those who maintained the ways of the spirit, the Children of the Law. This turn away from the spirit and this elevation of technology led to the destruction of the great continent. Though great power could be achieved through the use of Atlantean 'fire crystals,' great destruction was also always a threat. In the end the misuse of the crystals brought about the destruction of the civilization.

❝ MANY PEOPLE LIVING IN AMERICA IN THE TWENTIETH CENTURY WERE REINCARNATED ATLANTEANS ❞

Some of Cayce's followers also claim that he predicted the rediscovery of Atlantis – and that he gave clues to its whereabouts. According to an interpretation of one of his readings from 1927, Atlantis was to be uncovered off the southeastern coast of Florida, near the island of Bimini. In 1940 Cayce claimed in a reading that 'Poseidia will be among the first portions of Atlantis to rise again.'

Expect it in 68 or 69' (www.edgarcayce.com). The reference to 68 and 69 is taken to refer to the years 1968 and 1969. Some suggest that this is a prophetic utterance pointing toward the identification of Atlantis with the 'Bimini Road,' often said to have been discovered in 1968. The Bimini Road is half a mile of limestone rocks, laid out underwater in what appears to be a uniform pattern. To all intents and purposes it looks like an underwater road or the base of a huge wall. Robert Marx, writing in the November 1971 issue of *Argosy* magazine, speculated that the Bimini Road might be evidence of lost Atlantis; a man-made structure now covered with water. Though many geologists have insisted that the road is a purely natural formation, the facts seem all too clear to believers. Discovered in 1968, the Bimini Road fulfills Cayce's prophecy – Atlantis has begun to rise from the depths.

More recently, the Association for Research and Enlightenment, a group claiming association with the teaching of the late Edgar Cayce, has published claims concerning their search for Atlantis in the vicinity of Bimini. In addition to the Bimini Road, these researchers claim that they have discovered the remains of ancient harbors, now completely submerged, in the area. According to Greg Little's account of the organization's research in the July 2007 issue of *Alternate Perceptions Magazine*, expeditions in 2006 and 2007 were conducted using side-scan sonar. Sonar images indicated:

> rectangular formations lying in 100 feet of water off Bimini, several unusual stone formations 20+ miles out on the Great Bahama Bank, and the 'rediscovery' of an underwater 'mass' of fully dressed marble beams, an exquisite marble building apex, marble columns, and numerous huge, rectangular flat slabs of white marble. (http://www.mysterious-america.net/bermudatriangle0.html)

The claim that the remains of ancient Atlantis have been found in the area known as the 'Bermuda Triangle' has not escaped the notice of Atlantis researchers. This area of the Atlantic, stretching from Miami to the Bahamas to Bermuda, is also known as the 'Devil's Triangle' and is regarded as mysterious because of the supposedly large number of aircraft and ships that have disappeared without a trace within its waters. Theories of the origins of the Triangle's mysterious power have ranged from an underwater extraterrestrial base to the existence of a trans-dimensional portal in the area. A very prominent theory, however, is based on Cayce's location of Atlantis in the area and in his discussion of the Atlantean use of crystal power. According to this theory, planes and ships disappear in the Bermuda Triangle

16. Nazi Atlantean underwater gladiators from outer space
– *The Warlords of Atlantis* (1978). Don't mess with these guys.
Image courtesy of the Kobal Collection.

because the submerged Atlantean technology continues to operate, or at least continues to have an effect. The power of the Atlantean fire crystals is responsible for the disappearance of planes and ships.

The theory that lost Atlantis may be responsible for the mysteries of the Bermuda Triangle received support through the story of Dr. Ray Brown, who is said to have made a great discovery in the Bahamas in 1970. While diving, Brown became separated from his friends and discovered a large underwater pyramid. Exploring the capstone of the pyramid, Brown discovered an entranceway into its interior. In the center of a pyramidal room, Brown noted a brass rod descending from the ceiling. At its end was a many-faceted ruby-like gem. Below the gem was a stone plate that held a pair of sculpted metal hands. The hands held a three-and-a-half inch crystal. Brown removed the crystal and took it with him.

According to Jeffrey Keyte,

> Dr. Brown's crystal sphere has been the source of a wide variety of paranormal and mysterious occurrences. People have felt breezes or winds blowing close to it. Both cold and warm layers surround it at various distances. Other witnesses have observed phantom lights, heard voices or felt strange tingling sensations surrounding it.

A compass needle, when placed next to the crystal sphere, will spin counter-clockwise, then commence turning in the opposite direction when moved only inches away. Metals become temporarily magnetized when they come into close contact with the sphere. There are even recorded instances where healing has taken place by merely touching the sphere. (http://metareligion.com)

In 1976 Atlantean crystals came to the fore of the burgeoning 'New Age' movement. In that year Frank Alper began receiving channeled messages from ancient Atlantis that described the power of crystals, including the use of crystals for physical healings. Alper's claim was that crystals had the ability to store energy much like a battery and then release the energy in productive ways. Alper described the uses of Atlantean crystals in his three-volume *Exploring Atlantis*, the first volume of which was published in 1982.

According to Alper, the Atlanteans used giant fire crystals to power their cities. These crystals were located in underground tunnels and their energy was conducted through copper rods. Crystals were also used to amplify psychic abilities in order to allow telepathic communication between the Atlanteans. Pyramids in Atlantis were made of crystals and acted as antennae to receive and amplify universal energies and to establish magnetic energy grids that helped to hold the planet in its orbital path. A few of these pyramids still remain, but are buried deep underground. The locations of buried pyramids, including Mount Shasta, provide special healing power to those who visit them. Crystals were also used by the Atlanteans to treat pregnant women and to transfer information to their unborn fetuses. At two years of age, children left their parents to live in group homes, where crystals were used to transfer data directly to their brains. Alper argues that the healing power of crystals, known first to the Atlanteans, may be reclaimed by the human race today.

The Crystal Wings company of Mount Shasta, California sells, among many other crystals, 'The Lost Stone of Atlantis,' also known as 'The Channeling Stone of St. Germaine.' The business' website reports:

It seems that this 'missing stone' fills the gap and void of deep emotional trauma that we and our clients have been witnessing within, with a renewed purity of light and love. The energy of the stone itself transmutes and replaces the residual negativity and stagnancy that we hold in our heart center with a presence that many of us have not been familiar with for centuries upon centuries. Ruby Lavender Quartz™ dissolves one's blockages and in turn enables one to finally 'let go' and have the ability to manifest the true potentials of ones Kharmic calling. Without the walls, we become vessels for receiving and giving divine information and divine love. Ruby Lavender Quartz™ has truly become the embodiment of the unconditional mother, the unwavering nurturer, and the undaunted energy of love. It is the comfort of home. (www.crystalwings.com)

*   *   *

Okay, so I should have gone to Mount Shasta and purchased a Ruby Lavender Quartz crystal. 'Letting go' and 'manifesting the true potential of my Kharmic calling' sounds pretty good right now, not to mention 'replacing the residual negativity and stagnancy that I hold in my heart's center.' I'm sure that it would just as easily release me from the eternal circle of mud in which my soul now seems to be imprisoned.

But I'm here now, so I'll make the best of it.

A fellow spelunker gives me his hand and pulls me through the mud to solid ground. I am lucky to have my companions, members of the Little Rock Grotto, a club of Arkansas cave explorers, real-life adventurers, who agreed to be my guides through the depths of the cave. The group regularly explores caves throughout the region, mapping them and removing litter and debris. They have supplied me with knee pads, a helmet, and a water-proof bag, and lead me around boulders, push and pull me through openings that I would swear are smaller than the circumference of my head, warn me of sudden drops into cold water, and generally keep me from getting lost or killed.

The group has been here many times before, and knows something of the legend of the blue-skins and Sasquatch. Dewayne Agin, a member of the group, claims that he has never seen anything weird or out of the ordinary in this cave. Then he tells me about a fellow who used to explore this cave with their group. They nicknamed him 'Bigfoot' because he was full of stories of giant, hairy, forest people and unidentified flying objects.

Bigfoot was also known for exploring the cave alone and in the nude. That seems pretty weird to me.

*   *   *

What does it mean that Atlantis is lost? I don't mean lost beneath the sea, I mean lost to history, lost to science, lost to human knowledge. In its disappearance beneath the waves of the Atlantic, it offers a morality lesson – about the dangers of the thirst for power and an unhealthy dependence upon technology. Its disappearance from history provides a different lesson.

For Donnelly the loss was of pure religion, pure humanity.

For Blavatsky and the other theosophists, the loss was of humanity's true history – the loss of a history kind to both matter and the spirit.

For both, its loss was an illustration that humanity has lost its way. While Donnelly never brought this idea to clarity, Blavatsky and the others surely did. The message of the lost Atlantis is that history cannot be trusted, at least not the history of the mainstream, the history of the academy, the history of the schools. If we want to know the truth, we must look to other sources, we must look to the ascended masters, to mysterious ancient texts, to the esoteric sources known only to the few.

Herein lies the secret. While the Atlantis of Blavatsky, Scott-Elliot, Oliver and Cayce pretends to be the Atlantis of the chosen few, it is really the Atlantis of the masses. The knowledge of the theosophists is called esoteric, known only by the select few. But, in reality, the path to esoteric knowledge is wider than the path to accepted knowledge. Knowledge of Atlantis can be had for a nickel in cheap newsletters and pulp magazines. It can be found while waiting in the grocery store queue, on basic cable television channels, and on internet websites.

What does it mean that Atlantis is lost, lost to history and science? It means that we cannot trust those in authority, those in the ivory towers, those with credentials and honors and degrees.

Lost Atlantis is more than a utopian vision, meant to instruct or to warn.

Lost Atlantis, and lost Lemuria and Mu, are civilizations not so much from the pages of secret history as from the pages of rejected history, discredited history, sunken history, subterranean history. To believe in Atlantis is *not* to believe in the other story, the accepted story from textbooks and college lectures. To believe in Atlantis is to believe that wisdom can be found printed on pulp pages and that knowledge is often disguised as tabloid fodder.

FIVE

The Hollow Earth

I stand on the brink of a precipice. I cannot see the bottom. I grasp the rope in front of me with both hands. I rest my weight against the rope to my back. I inch my feet along the narrow ledge, trying not to think about what I am doing.

I am Tom Sawyer in the treasure cave. I am Frodo in the Mines of Moria. I am David Innes in Pellucidar.

This is the story that has led me into the caverns of the Ozarks.

It was in June of 1964 that a group of experienced cavers returned to Blowing Cave near Cushman, Arkansas for what they imagined to be just one of many exciting but uneventful expeditions into the cave. On this occasion, however, the unexpected happened. While exploring a portion of the cave floor that was covered with debris, the men noticed an opening under one of the rocks that seemed to open wide enough for a descent. After moving some of the heavy stones they began the climb down into a previously unexplored part of the cave. It was likely that they had walked over or around this opening countless times in their many trips to Blowing Cave, little realizing the wonders to be found just under their feet.

After traveling down for almost four miles, the explorers found themselves in an enormous cavern, which they named Glass Cave. In the floor of Glass Cave, the men discovered yet another crevice partially hidden under rocks. Reaching the bottom of the crevice they found that they were in a large, smooth-walled tunnel. With a domed ceiling and flat floor, the tunnel clearly looked to be man made. It was illuminated by a faint green glow that emanated from the walls, which were clear like glass. In some areas they could see through the transparent walls into the cavern around them and could make out the faint shapes of moving creatures – giant serpents, reptiles, and insects.

On the third day of their journey they encountered a race of tall, blue-skinned people. Using an electronic device, the Blues were able to communicate with their visitors. The surface dwellers learned that the Blues were representatives of a great subterranean civilization, whose origins predated those of surface civilizations by thousands of years, perhaps stretching back to the ancient days of Atlantis. They shared the deep caverns with the giant insects and reptiles and with a race of hairy Sasquatch-like creatures. These creatures were exceptionally violent and seemed to be engaged in an eternal struggle with the giant reptiles. The Sasquatch had the advantage, however, as they were armed for their conflicts with laser guns.

While the rest of the group returned to the surface world, at least one of their members stayed behind.

Now I am in Blowing Cave, led by an experienced group of cavers from Little Rock, Arkansas.

I am trying hard not to think of the fate of the cavers from the movie The Descent, *but that is proving hard to do. Pale-skinned, blind, cannibalistic troglodytes push the friendly blue-skins out of my mind.*

EDMUND HALLEY

In 1692 Edmund Halley, best known for his discovery of the ever-returning comet that bears his name, offered up a theory of the earth's inner structure that was meant to solve the problem of the motion of the magnetic poles. His solution to the problem was quite ingenious. We must suppose, he said, that the earth is composed of three parts: the outer shell with which we are so familiar, an inner globe or nucleus, and a fluid medium in between. If we further imagine that both the inner and the outer globe are turning around a common center and axis of rotation but slightly out of sync with one another, that is with one turning slightly faster than the other, we can discover a solution to the problem of the motion of the magnetic poles (Fitting, 20).

Halley's theory did not stop there, however. He also suggests that, since life has been observed to thrive in all parts of creation, we might suppose that the inner globe is itself populated with animate life forms. Halley wrote in his *A Theory of Magnetic Variations*:

> But since we see all the parts of the Creation abound with animate Beings, why should we think it strange that the prodigious Mass of Matter, whereof this Globe does consist, should be capable of some

other Improvements, than barely to serve to support its Surface? Why may we not rather suppose that the exceeding small Quantity of solid Matter in respect to the fluid Aether, is so disposed by the Almighty Wisdom, as to yield as great a Surface for the use of Living Creatures, as can consist with the Conveniency and Security of the whole? (quoted in Fitting, 22)

It might further be supposed, Halley wrote, that the inner, concave surface of the outer shell may glow with sufficient light to provide warmth and vision to the creatures living upon the inner surface. Halley's theory also offered an explanation for the mysterious polar lights, explained as the leaking of some of the inner fluid medium into our atmosphere at the poles, where the surface of the outer shell is thinner than at other locations.

Though Halley's scientific theories would not remain in favor among the scientific establishment, they did prove to be quite long-lived in other communities. Halley's theories may indeed be seen as a bridge between earlier religious or folkloric accounts of hell and subterranean gnomes and later science fiction accounts of the hollow

17. 1951's *Superman and the Mole-Men*. See Superman battle the tiny vacuum-cleaner salesmen from the center of the Earth!

earth, like those created by Jules Verne and Edgar Rice Burroughs. They may also be seen as a bridge to a more fully developed, if not more fully persuasive, set of hollow earth theories.

Symmes' Holes

John Cleve Symmes, Jr. of St. Louis, Missouri issued a circular to 'each notable foreign government, reigning prince, legislature, city, college, and philosophical society, quite around the earth' (Fitting, 95). According to Fitting, the statement was accompanied by a testament to the author's sanity. The circular, published in 1818, read:

To All the World!

I declare the world is hollow, and habitable within; containing a number of solid concentric spheres, one within the other, and that it is open at the poles twelve or sixteen degrees; I pledge my life in support of this truth, and am ready to explore the hollow, if the world will support and aid me in the undertaking . . .

> ❝ I DECLARE THE WORLD IS HOLLOW, AND HABITABLE WITHIN ❞

I ask one hundred brave companions, well equipped, to start from Siberia in the fall season, with Reindeer and sleighs, on the ice of the frozen sea; I engage we find a warm and rich land, stocked with thrifty vegetables and animals, if not men, on reaching one degree northward of latitude 82; we will return in the succeeding spring. (95)

Though Symmes himself never elaborated at great length upon his theory of a hollow earth and polar openings, some of his supporters did, most notably James McBride in *Symmes's Theory of Concentric Spheres, Demonstrating that the Earth is Hollow, Habitable Within, and Widely Open about the Poles*, of 1826. McBride indicates that Symmes' theory was meant to be inclusive of all celestial bodies, from the sun to meteors. The theory claims that each of these objects is composed of solid concentric spheres that are open at the poles. These spheres are separated from one another by a space filled with aerial fluids. The earth is composed of five hollow concentric spheres, with the polar openings measuring between 4,000 and 6,000 miles in diameter.

According to McBride's reading of Symmes' theory,

Each of the spheres composing the earth, as well as those constituting the other planets throughout the universe, is believed to habitable both on the inner and outer surface; and lighted and warmed according to those general laws which communicate light and heat to every part of the universe. (Fitting, 109)

In 1820, just on the heels of the publication of Symmes' initial circular, a novelized version of Symmes' theory appeared under the title *Symzonia*. Though often attributed to Symmes, the authorship of *Symzonia* remains in doubt. The novel tells the story of a journey by ship through the northern polar opening into the inhabited world of the earth's interior. Interestingly, the world of *Symzonia* is not described as existing on the surface of an interior sphere, but rather on the concave surface of the earth's outer shell, a position which Symmes himself may have adopted at a later period.

After passing through the Arctic Circle into a tropically warm polar climate, the adventurers enter the hollow earth. The inhabitants of the earth's core are humanoid and described as gentle vegetarians. They reject material possessions in favor of the wealth of the soul. Because of their virtuous nature, the internals possess great intelligence and have little need for sleep. They are extremely attractive, with beautiful white skin. They also have great physical strength. The government is democratic and women are granted equal status with men. In many ways they seem to exist in a Garden of Eden, like humanity before the fall from grace.

Despite their lack of desire for material possessions, the internals had a highly developed technology, including air-ships. The narrator, John Seaborn, describes his encounter with these amazing craft:

I had not been long at my study of language, when Mr. Albicore sent me word that a bird as big as the ship was coming towards us. I went on deck, and immediately saw that Albicore's bird was no other than an aerial vessel, with a number of men on board. It came directly over the ship, and descended so low that the people in it spoke with the internals who were with me; but I was not yet qualified to understand a word of what passed. I observed its appearance to be that of a ship's barge, with an inflated wind sail, in the form of a cylinder, suspended longitudinally over it, leaving a space in which were the people. It had a rudder like a fishes tail, and fins or oars, which appeared to be moved by the people within. On the whole it was not a matter of great surprise to me. I only inferred from it, that the internals understood aerostatics much better than the externals. (Fitting, 199–200)

Seaborn later learned that the inflated cylinder above the boat was filled with an 'elastic gas' that diminished the specific gravity of the craft and allowed it to fly.

Of course, not everything is perfect, even in utopian Symzonia. Occasionally evils are committed and the evildoers are punished by being exiled to the outer world. Indeed, those of us who live on the outer surface of our planet are descended from the exiled criminals of Symzonia. The surface of the earth, like Australia, was populated by society's convicts and degenerates. After reading surface literature supplied to them by Seaborn, the internals decide that the presence of this degenerate race cannot be allowed to influence their society. Seaborn and his companions are ordered to leave and to return to the surface, which they do, sailing on the internal southern sea, through the southern polar opening, and home.

KORESHANITY

Symmes' earth theory was not the only such theory from the nineteenth century that now strikes most of us as a little odd. Equally unacceptable to today's cosmology is the version of reality proposed by utopian millenarian Cyrus Teed. Teed claimed that in 1869 he experienced an epiphany in which his true identity as the reincarnation of Christ was revealed to him. This revelation concerning his own nature also came with a new understanding of the divine nature as possessed of both male and female qualities, and of human nature, including a belief in the equity of men and women and the need for society to transform itself accordingly into a community of shared wealth. Teed took the name 'Koresh' and attracted a small group of followers who adopted his method of communal living and accepted Koresh as the living incarnation of God. The commune began in Chicago and then moved to Florida, where they established a colony on the banks of the Estero River. While Teed's ideas about communal living, equity for men and women, the coming 'end times,' and the nature of God fit the pattern of many millenarian groups of the time, he did possess some truly unique ideas in regards to the nature of the cosmos.

Cyrus Teed believed in the hollow earth, though his ideas about humanity's place in regards to the hollow world were decidedly different from Symmes'. The Symmes version argued that the earth is a hollow ball with humanity living upon its outer surface and the interior accessible by holes at the north and south poles. Teed agreed with the basic picture of the earth as a hollow ball. The difference was that he believed that *we* reside on the inside surface of the ball – on the concave surface rather than the convex surface. The heavens, including the sun, the moon, and the stars, are located on the convex surface of a smaller globe inside the earth. From any point on the inner surface of the larger sphere, where we reside, we look up, or toward the center of the earth, to see the heavens. For Teed, the acceptance of this view of cosmology was quite important, having spiritual as well as empirical value. He wrote in *The Cellular Cosmogony* of 1898 that to know of the earth's concavity and its relation to universal form is to know God; while to believe in the earth's convexity is to deny him and all his works. All that is opposed to Koreshanity is antichrist (84).

Koresh believed that all of creation possessed the same basic structure, the cellular structure of an egg. The sun is the center, the earth is the shell, and we live on the shell's *inner* surface. Koresh spends page after page of this book explaining how such things as the movements of the sun and moon, the movement of the stars, and solar and lunar eclipses can be made sense of within his system.

Though the Koreshan system failed to appeal to a larger audience and never achieved even the limited number of supporters that Symmes' theory attracted, it is rumored to have had at least one fervent believer. Legend has it that Adolf Hitler became a convert to Koreshanity and attempted to use his newly acquired knowledge to assist in his program of war. Supposedly, Hitler had his scientists calculate the exact location from which it would be possible to look straight up and observe, with the use of powerful telescopes, the movements of the British navy.

THE PHANTOM OF THE POLES

In 1906, William Reed's *The Phantom of the Poles* revived the hollow earth theory for a world that was growing obsessed with the exploration of the polar regions. Reed argued that reaching the North Pole was an impossibility because it did not exist. Those intent on reaching the pole were engaged in chasing a phantom. If one continued in a northward direction, one would not reach the pole but would rather travel over the curve that leads into the hollow interior. Indeed, Reed claims that many arctic explorers have no doubt already traveled into the interior without recognizing it.

Reed's book seeks to build a case for a hollow earth, open at both ends, by examining a series of phenomena in the light of his theory. For example, he notes that it is generally recognized that the earth is flattened on the ends. Why is this the case? 'As the earth is hollow, it could not be round, is the answer to that. Again, the opening to the interior would detract from its roundness just in proportion to the size of the opening' (21). Can his theory explain the strange lights of the Aurora Borealis, a mystery that the science of his day found quite perplexing?

> The Aurora Borealis is the reflection of a fire within the interior of the earth. The exploding and igniting of a burning volcano, containing all kinds of minerals, oils, and so on, causes much coloring; while absence of coloring, or only a faint toning, is due to the burning of vegetable matter, such as prairie or forest fires. (23)

How do icebergs of freshwater arise in the midst of a salty ocean?

> Where are they formed? And how? In the interior of the earth, where it is warm, by streams or canyons flowing to the Arctic Circle, where it is very cold, the mouth of the stream freezing and the water, continuing to pass over it, freezing as it flows. This prevails for months, until, owing to the warm weather in summer, the warmth from the earth, and the warm rains passing down to the sea, the bergs are thawed loose and washed into the ocean. Icebergs cannot be formed on earth, for the reason that it is colder inland than at the mouth of a stream; hence the mouth would be the last to freeze and the first to thaw. Under those conditions, icebergs could not be formed. (24)

Of course, the biggest question of the day for those interested in the exploration of the poles was: why had no one been able to reach it?

> The curve leading to the interior of the earth may be land or water, just as it happens, and he that passes to the farthest point of the circle when the observation is taken, will show the farthest point north. But if he continues straight on he will soon be losing ground, or getting farther from the supposed pole, and eventually be going south and not know it, as the compass could not then be depended upon. There must be some good reason. (272)

Seeking the North Pole, he writes, is like seeking to reach the moon by chasing its reflection in a body of water. As you move, the moon moves with you. The moon in the water is only a phantom, as is the pole. To continue northward is to continue over the curve of the earth and into the inner world.

THE SMOKEY GOD

The 1908 account *The Smokey God, or A Voyage to the Inner World* by Willis George Emerson tells the story of Olaf Jansen. Emerson describes Jansen as an elderly Norwegian of 95 who was residing in southern California when they met and where Jansen related his deathbed story. Having spent years in a mental institution for telling his tale in the past, he had vowed to keep it secret until he was near to death. The story was of a fishing trip with his father when he was 19 years old.

Jansen's story begins like earlier tales and tells of sailing far north toward the pole only to find the climate warm and tropical. After sailing for weeks, the Jansens finally came into sight of land. Sailing up a river toward the interior of the continent they caught sight of a huge craft. The passengers were singing and playing harps. When the craft passed, a smaller boat was lowered to the water with six gigantic men aboard. The visitors invited Jansen and his father aboard their craft, an invitation that they readily accepted. Jansen described these strange giants:

> There was not a single man aboard who would not have measured fully twelve feet in height. They all wore full beards, not particularly long, but seemingly short-cropped. They had mild and beautiful faces, exceeding fair, with ruddy complexions. The hair and beard of some were black, others sandy, and still others yellow. The captain, as we designated the dignitary in command of the great vessel, was fully a head taller than any of his companions. The women averaged from ten to eleven feet in height. Their features were especially regular and refined, while their complexion was of a most delicate tint heightened by a healthful glow. (98–9)

As the ship sped back up river, Jansen noted the warmth of the strange sun over their heads. It glowed with a white light from a large bank of cloud. The refraction of the sun's light through the clouds gave the light of two full moons on a clear night. This cloudy sun was found to shine for 12 hours and then disappear, as if eclipsed. Jansen learned that his hosts worshipped this light, 'The Smokey God.' As

❝ THEY HAD SAILED OVER THE CURVE OF THE EARTH'S OUTER SHELL AND INTO THE HOLLOW INTERIOR ❞ Jansen and his father learned that they had sailed over the curve of the earth's outer shell and into the hollow interior along the inner surface of the shell, they came to understand that the sun now shining above their heads was a glowing ball of electricity at the earth's core. From the perspective of inhabitants of the outer surface, the inner surface looked down upon the sun.

The Jansens journeyed to the beautiful city known as Eden by a kind of train, and experienced many more great wonders, including a great herd of elephants:

> There must have been five hundred of these thunder-throated monsters, with their restlessly waving trunks. They were tearing huge boughs from the trees and trampling smaller growth into dust like so much hazel-brush. They would average over 100 feet in length and from 75 to 85 in height.
>
> It seemed as I gazed upon this wonderful herd of giant elephants, that I was again living in the public library at Stockholm, where I had spent much time studying the wonders of the Miocene age. I was filled with mute astonishment, and my father was speechless with awe. He held my arm with a protecting grip, as if fearful harm would overtake us. We were two atoms in this great forest, and fortunately, unobserved by this vast herd of elephants as they drifted on and away, following a leader as does a herd of sheep. They browsed from growing herbage which they encountered as they traveled, and now and again shook the firmament with their deep bellowing. (126–7)

Finally, their adventures over, Olaf and his father sailed across the inner ocean and around the curve of the earth's southern polar opening.

A JOURNEY TO THE EARTH'S INTERIOR

Seven years after Reed's *Phantom of the Poles*, which sought to show why the ongoing quest to reach the North Pole was bound for failure, Marshall Gardner's 1913 *A Journey to the Earth's Interior* faced a slightly more difficult obstacle when it came to persuading readers that reaching the pole was an impossibility. The obstacle? The recent claims by both Cook and Peary to have done just that. Gardner not only had to marshal evidence to support his hollow earth theory, he also had to demonstrate why he believed that Cook's and Peary's claims were mistaken. He also, it seems, wanted to demonstrate the superiority of his arguments over those that had come before. He begins, therefore, with an important caveat aimed at Symmes and perhaps Reed as well.

Cranky Ideas are Not in Same Class with Scientific Ones

It will also be an injustice to us if the reader confuses our idea of a hollow earth as presented in this book with one or two theories which have been put out in the past and which only bear a superficial relation to ours. For instance, nearly one hundred years ago in America a theory was put forth that the earth consisted of a number of concentric spheres one within the other. Now that could hardly be called a scientific theory. It was based on a supposition, and the author argued from his supposition down to what the facts ought to be. He said in effect, 'According to my principle there ought to be within the earth a series of spheres each one inside the other'. But he did not know, and he never went down to see. (27)

Gardner offers a diversity of evidence to support his claim, including evidence from the field of astronomy. He argues that the fact that other planets are hollow spheres lends credence to the idea that the earth is hollow as well. Examining Mars and Venus through a telescope, he claims, reveals that they are indeed hollow. What some researchers have mistaken for polar ice caps on the planet Mars are, in reality, holes leading into the interior. On certain occasions the glimmer of an interior sun has even been glimpsed.

In addition, Gardner claims that the discovery of the preserved remains of woolly mammoths can best be explained by the hollow earth theory. The creatures obviously continue to thrive within the earth's core and occasionally wander through the polar opening and onto the surface of the earth. As they travel through the Arctic Circle they become frozen in ice until discovered by humans a short time later.

Of course, his evidence means little unless he can show why the claims of Peary and Cook should be disregarded. How is it, readers may wonder, that Peary does not describe the huge opening at the North Pole in his account of his polar expedition. The answer is quite simple – the hole into the interior is not something that one would expect to notice from a position on the surface of the earth. It is the same reason, he argues, that humans did not realize that the earth was round:

> Why did not man discover by looking around him, that he was living on the surface of what is, practically speaking, an immense sphere (to be exact spheroid)? And why did man for centuries think that the earth was flat? Simply because the sphere was so large that he could not see its curvature but thought it was a flat surface, and that he should be able to move all over the surface of it appeared so natural that, when scientists first told him it was a sphere he began to wonder why he did not fall off, or at least, if he lived in the northern hemisphere, he wondered why the Australians did not fall off – for he had no conception of the law of gravity.

> Now, in the case of the polar explorers the same thing is true. They sail up to the outer edge of the immense polar opening, but that opening is so vast – remember that the crust of the earth over which it curves is eight hundred miles thick – that the down-ward curvature of its edge is not perceptible to them, and its diameter is so great – say 1400 miles – that its other side is not visible to them. So that if an explorer went far enough he could sail right over that edge, down over the seas of the inner world and out through the

18. The Iron Mole that transports Doug McClure and Peter Cushing to the land of Pellucidar in *At the Earth's Core* (1976).
Image courtesy of the Kobal Collection.

19. Caroline Munro as a beautiful troglodyte – *At the Earth's Core* (1976). Image courtesy of the Kobal Collection.

Antarctic orifice, and all that would show him what he had done, would be that as soon as he got inside he would see a smaller sun than he was accustomed to – only to him it might look larger owing to its closeness – and he would not be able to take any observations by the stars because there would be neither stars nor even a night in which to see them. (33–4)

Neither Peary nor Cook's brief excursions to the polar north were prepared to investigate the veracity of the hollow earth theory. Believing the earth to be a solid sphere, they would have missed the evidence for the polar opening entirely. It is only by examining the evidence with an open mind that the truth can be discovered.

The Lost Diary of Admiral Byrd

Raymond Bernard's *The Hollow Earth* (1964) covered much of the same territory that had already been staked out by Reed and Gardner, including chapters devoted to each of their books. His work did more than simply rework those earlier texts, however, for Bernard's book makes some important contributions of its own. Bernard would link the hollow earth theory with unidentified flying objects and claim that flying saucers were not from outer space but from the earth's core; he would introduce claims concerning Admiral Byrd's support for the hollow earth theory; and he would connect the Symmes/Reed/Gardner tradition of hollow earth theory with other strands of esoterica.

Bernard began his book with a dedication that seemed to make the discovery of the interior world an event that was just around the corner, just waiting for a brave adventurer to answer the call. He wrote:

Dedicated to the Future Explorers of the New World that exists beyond North and South Poles in the hollow interior of the Earth. Who will Repeat Admiral Byrd's historic flight for 1,700 miles beyond the North Pole and that of his Expedition for 2,300 Miles beyond the South Pole, entering a New Unknown Territory not shown on any map, covering an immense land area whose total size is larger than North America, consisting of forests, mountains, lakes, vegetation and animal life.

The aviator who will be the first to reach this New Territory, unknown until Admiral Byrd first discovered it, will go down in history as a New Columbus and greater than Columbus, for while Columbus discovered a new continent, he will discover a new world. (5)

Like the other writers in this genre, Bernard's goals are quite lofty, and he offers the requisite laundry list of things that he seeks to prove. Some of them are familiar: the earth is hollow with polar openings, the poles have never been reached as this is an impossibility, and the hollow earth is probably filled with plant and animal life waiting to be discovered. Some of them are new: it is more logical to suppose that flying saucers come from the earth's interior than from another planet, the inner earth will provide a safe haven for refugees from a nuclear holocaust, and:

> That the observations and discoveries of Rear Admiral Richard E. Byrd of the United States Navy, who was the first to enter into the polar openings, which he did for a total distance of 4,000 miles in the Arctic and Antarctic, confirm the correctness of our revolutionary theory of the Earth's structure, as do the observations of other Arctic explorers. (21)

Admiral Byrd, according to Bernard, made several interesting comments concerning his flights to the North and South Poles, which indicate that he had indeed crossed over into the earth's interior. He quotes Byrd as saying in February, 1947, 'I'd like to see that *land beyond the* (North) *Pole*. That area beyond the Pole is the center of the *Great Unknown*,' and in March, 1956, following his return from the South Pole, 'The present expedition has opened up a *vast new territory*' (20). Bernard also notes that Byrd referred to the land beyond the poles as 'that enchanted continent in the sky, *land of everlasting mystery!*'

❛ I'D LIKE TO SEE THAT LAND BEYOND THE NORTH POLE. THAT AREA BEYOND THE POLE IS THE CENTER OF THE GREAT UNKNOWN ❜

Bernard writes:

> The only way that we can understand Byrd's enigmatical statements is if we discard the traditional conception of the formation of the earth and entertain an entirely new one, according to which its Arctic and Antarctic extremities are not convex but concave, and that Byrd entered into the polar concavities when he went beyond the Poles.

> In other words, he did not travel across the Poles to the other side, but entered into the polar concavity or depression, which . . . opens to the hollow interior of the earth, the home of plant, animal and human life, enjoying a tropical climate. This is the 'Great Unknown' to which Byrd had reference when he made this statement – and not the ice – and snow-bound area on the other side of the North Pole, extending to the upper reaches of Siberia. (30)

Likewise:

> The expression 'that enchanted continent in the sky' obviously refers to a land area, and not ice, mirrored in the sky which acts as a mirror, a strange phenomenon observed by many polar explorers, who speak of 'the island in the sky' or 'water sky,' depending or whether the sky of polar regions reflects land or water. If Byrd saw the reflection of water or ice he would not use the word 'continent,' nor call it an 'enchanted' continent. It was 'enchanted' because, according to accepted geographical conceptions, this continent which Byrd saw reflected in the sky (where water globules act as a mirror for the surface below) could not exist. (34–5)

Adding further detail to Bernard's claims is Byrd's strange 'missing diary' whose origins remain mysterious but which has been passed around in hollow earth circles, sometimes in cheap mimeographed versions, for years.

Byrd's missing diary begins:

> I must write this diary in secrecy and obscurity. It concerns my Arctic flight of the nineteenth day of February in the year of Nineteen and Forty Seven.
>   There comes a time when the rationality of men must fade into insignificance and one must accept the inevitability of the Truth! I am not at liberty to disclose the following documentation at this writing . . . perhaps it shall never see the light of public scrutiny, but I must do my duty and record here for all to read one day. In a world of greed and exploitation of certain of mankind, one can no longer suppress that which is truth.

Following this rather sensational introduction, the diary takes on the form of a flight log. In some versions that I have seen the hours are recorded, in others they are missing. I will include them here as they add to the drama.

0600 HOURS – All preparations are complete for our flight northward and we are airborne with full fuel tanks at 0610 Hours.

0620 HOURS – Fuel mixture on starboard engine seems too rich, adjustment made and Pratt Whittneys are running smoothly.

0730 HOURS – Radio Check with base camp. All is well and radio reception is normal . . .

0910 HOURS – Both Magnetic and Gyro compasses beginning to gyrate and wobble, we are unable to hold our heading by instrumentation. Take bearing with Sun compass, yet all seems well. The controls are seemingly slow to respond and have sluggish quality, but there is no indication of Icing!

0915 HOURS – In the distance is what appears to be mountains.

0949 HOURS – 29 minutes elapsed flight time from the first sighting of the mountains, it is no illusion. They are mountains and consisting of a small range that I have never seen before! . . .

1000 HOURS – We are crossing over the small mountain range and still proceeding northward as best as can be ascertained. Beyond the mountain range is what appears to be a valley with a small river or stream running through the center portion. There should be no green valley below! Something is definitely wrong and abnormal here! We should be over Ice and Snow! To the portside are great forests growing on the mountain slopes. Our navigation Instruments are still spinning, the gyroscope is oscillating back and forth!

1005 HOURS – I alter altitude to 1400 feet and execute a sharp left turn to better examine the valley below. It is green with either moss or a type of tight knit grass. The Light here seems different. I cannot see the Sun anymore. We make another left turn and we spot what seems to be a large animal of some kind below us. It appears to be an elephant! NO!!! It looks more like a mammoth! This is incredible! Yet, there it is! Decrease altitude to 1000 feet and take binoculars to better examine the animal. It is confirmed – it is definitely a mammoth-like animal! Report this to base camp . . .

1130 HOURS – Countryside below is more level and normal (if I may use that word). Ahead we spot what seems to be a city! This is impossible! Aircraft seems light and oddly buoyant. The controls refuse to respond! My GOD! Off our port and starboard wings are a strange type of aircraft. They are closing rapidly alongside! They are disc-shaped and have a radiant quality to them. They are close enough now to see the markings on them. It is a type of Swastika!!! This is fantastic. Where are we! What has happened. I tug at the controls again. They will not respond! We are caught in an invisible vice grip of some type!

1135 HOURS – Our radio crackles and a voice comes through in English with what perhaps is a slight Nordic or Germanic accent! The message is: 'Welcome, Admiral, to our domain. We shall land you in exactly seven minutes! Relax, Admiral, you are in good hands.' I note the engines of our plane have stopped running! The aircraft is under some strange control and is now turning itself. The controls are useless.

1140 HOURS – Another radio message received. We begin the landing process now, and in moments the plane shudders slightly, and begins a descent as though caught in some great unseen elevator! The downward motion is negligible, and we touch down with only a slight jolt!

1145 HOURS – I am making a hasty last entry in the flight log. Several men are approaching on foot toward our aircraft. They are tall with blond hair. In the distance is a large shimmering city pulsating with rainbow hues of color. I do not know what is going to happen now, but I see no signs of weapons on those approaching. I hear now a voice ordering me by name to open the cargo door. I comply.

From this point the diary offers Byrd's recollections of the events following the landing. Byrd is soon taken from his aircraft to a city made of crystal and to a building 'that is a type I have never seen before. It appears to be right out of the design board of Frank Lloyd Wright, or perhaps more correctly, out of a Buck Rogers setting!!' (37). Once inside the building, Byrd is ushered in to meet a spiritual master. Byrd recalls the master's words and his own response to them:

'We have let you enter here because you are of noble character and well-known on the Surface World, Admiral.' Surface World, I half-gasp under my breath! 'Yes,' the Master replies with a smile, 'you are in the domain of the Arianni, the Inner World of the Earth. We shall not long delay your mission, and you will be safely escorted back to the surface and for a distance beyond. But now, Admiral, I shall tell you why you have been summoned here. Our interest rightly begins just after your race exploded the first atomic bombs over Hiroshima and Nagasaki, Japan. It was at that alarming time we sent our flying machines, the "Flugelrads," to your surface world to investigate what your race had done. That is, of course, past history now, my dear Admiral, but I must continue on.

'You see, we have never interfered before in your race's wars, and barbarity, but now we must, for you have learned to tamper with a certain power that is not for man, namely, that of atomic energy. Our emissaries have already delivered messages to the powers of your world, and yet they do not heed. Now you have been chosen to be witness here that our world does exist. You see, our Culture and Science is many thousands of years beyond your race, Admiral.' (40–2)

After the master has employed Byrd to warn the world of the impending nuclear disaster and to share the story of the inner world, he is ushered back to his aircraft. The flight log continues the story:

215 HOURS – A radio message comes through. 'We are leaving you now, Admiral, your controls are free. Auf Wiedersehen!' We watched for a moment as the flugelrads disappeared into the pale blue sky. The aircraft suddenly felt as though caught in a sharp downdraft for a moment. We quickly recovered her control. We do not speak for some time, each man has his thoughts . . .

300 HOURS – We land smoothly at base camp. I have a mission . . .

Then follows an entry from March 11, 1947:

MARCH 11, 1947. I have just attended a staff meeting at the Pentagon. I have stated fully my discovery and the message from the Master. All is duly recorded. The President has been advised. I am now detained for several hours (six hours, thirty-nine minutes, to be exact.) I am interviewed intently by Top Security Forces and a medical team. It was an ordeal! I am placed under strict control via the national security provisions of this United States of America. I am ORDERED TO REMAIN SILENT IN REGARD TO ALL THAT I HAVE LEARNED, ON THE BEHALF OF HUMANITY! Incredible! I am reminded that I am a military man and I must obey orders. (59–60)

And, the final entry from December 30, 1956:

These last few years elapsed since 1947 have not been kind . . . I now make my final entry in this singular diary. In closing, I must state that I have faithfully kept this matter secret as directed all these years. It has been completely against my values of moral right. Now, I seem to sense the long night coming on and this secret will not die with me, but as all truth shall, it will triumph and so it shall.

This can be the only hope for mankind. I have seen the truth and it has quickened my spirit and has set me free! I have done my duty toward the monstrous military industrial complex. Now, the long night begins to approach, but there shall be no end. Just as the long night of the Arctic ends, the brilliant sunshine of Truth shall come again . . . and those who are of darkness shall fall in its Light. FOR I HAVE SEEN THAT LAND BEYOND THE POLE, THAT CENTER OF THE GREAT UNKNOWN.

The Byrd diary, like Bernard's *Hollow Earth*, freely combines elements of the Symmes/Reed/Gardner hollow earth tradition with the traditions of Atlantis and Lemuria as articulated by the theosophists. Bernard, for example, describes the civilization at the center of the earth in terms of Agharta and Shambala, which are, according to Bernard, names given to the underground world by Tibetan Buddhists who believe that the King of the World rules from below ground. This underground kingdom of Agharta, ruled over by the capital city of Shambala, is far more advanced than our own, Bernard claims, and is responsible for the flying saucers which fill our skies. Tunnels connect this underground kingdom with all areas of the world, making it accessible from many places other than the poles. Descended from the survivors of the sunken continents of Lemuria and Atlantis, the kingdom at the center of the earth has much to teach us, both technologically and spiritually.

VRIL

Preceding Jules Verne's subterranean novel by just a few years, Edward Bulwer-Lytton's *Vril: The Power of the Coming Race* tells the story of the narrator's descent through a mine shaft into the cavern home of the Vril-ya, a race of humans who fled from floods on the earth's surface in ancient times, lost their way, and established a civilization in underground caverns. This account places the subterranean civilization in caverns beneath the earth's surface rather than in the earth's hollow interior.

The narrator describes a prosperous and advanced civilization with fields of vegetables and running water. Canals and buildings have been carved out of the stone walls of the caverns. And the residents?

And now there came out of this building a form – human: – was it human? It stood on the broad way and looked around, beheld me and approached. It came within a few yards of me, and at the sight and presence of it an indescribable awe and tremor seized me, rooting my feet to the ground . . . It was tall, not gigantic, but tall as the tallest men below the height of giants.

Its chief covering seemed to me to be composed of large wings folded over its breast and reaching to its knees; the rest of its attire was composed of an under tunic and leggings of some thin fibrous material. It wore on its head a kind of tiara that shone with jewels, and carried in its right hand a slender staff of bright metal like polished steel. But the face! it was that which inspired my awe and my terror. It was the face of a man, but yet of a type of man distinct from our known extant races. The nearest approach to it in outline and expression is the face of the sculptured sphinx – so regular in its calm, intellectual, mysterious beauty. (19–20)

❝ IT WORE ON ITS HEAD A KIND OF TIARA THAT SHONE . . . AND CARRIED IN ITS HAND A SLENDER STAFF OF BRIGHT METAL ❞

The wings of the stranger turned out to be an apparatus rather than natural appendages, but, like the wings of a bird, they allowed the residents of the inner earth to fly throughout the caverns.

The Vril-ya quickly learned English from their visitor and shared with him the secret of their ability to fly and their other remarkable feats. Their great wonders, he was told, were accomplished through the use of vril – a source of power that pervades all of the universe. The narrator describes vril in this way:

I should call it electricity, except that it comprehends in its manifold branches other forces of nature, to which, in our scientific nomenclature, differing names are assigned, such as magnetism, &c. These people consider that in vril they have arrived at the unity in natural energic agencies, which have been conjectured by many philosophers above ground, and which Faraday thus intimates under the more cautious term of correlation. (45)

By mastering the vril force, which they do through sheer will-power, the Vril-ya are able to accomplish amazing things, including influencing the weather. The vril power allows its users to influence other human minds and bodies: 'The faculties of the mind could be quickened to a degree unknown in the waking state, by trance or vision, in which the thoughts of one brain could be transmitted to another, and knowledge be thus rapidly interchanged' (46). Vril can be used to carve through solid rock, a technique used by the Vril-ya to create their homes and buildings. Vril is the source of subterranean light and heat. In addition to these seemingly beneficial uses of vril, however, it is also revealed that a wielder of such great power can also do great harm. Vril can be used to destroy as well as to heal and teach:

> The fire lodged in the hollow of a rod directed by the hand of a child could shatter the strongest fortress, or cleave its burning way from the van to the rear of an embattled host. If army met army, and both had command of this agency, it could be but to the annihilation of each. (56)

The rod wielded by the narrator's hypothetical child is otherwise referred to as the Vril Staff and is described in some detail:

> It is hollow, and has in the handle several stops, keys, or springs by which its force can be altered, modified, or directed – so that by one process it destroys, by another it heals – by one it can rend the rock, by another disperse the vapour – by one it affects bodies, by another it can exercise a certain influence over minds. It is usually carried in the convenient size of a walking-staff, but it has slides by which it can be lengthened or shortened at will. (111)

It is this power of destruction rather than of healing and progress that resides with the narrator after he returns to the surface world. He wonders what might happen on that day when the Vril-ya decide to come to the surface. He muses:

> The more I think of a people calmly developing, in regions excluded from our sight and deemed uninhabitable by our sages, powers surpassing our most disciplined modes of force, and virtues to which our life, social and political, becomes antagonistic in proportion as our civilization advances, the more devoutly I pray that ages may yet elapse before there emerge into sunlight our inevitable destroyers . . . I have thought it my duty to my fellow-men to place on record these forewarnings of The Coming Race. (248)

Far from being taken as an early example of science fiction, Bulwer-Lytton's novel has been taken, like *The Smokey God*, as more

THE HOLLOW EARTH * 151

than a work of fiction. For example, the concept of vril seems to have had some influence upon Blavatsky who wrote in *Isis Unveiled*: 'Sir E. Bulwer-Lytton, in his *Coming Race*, describes it as the VRIL, used by the subterranean populations, and allowed his readers to take it for a fiction' (126). Alec Maclellan claims that the influence is far greater than this mention by Blavatsky would indicate, however. In his account of the hollow earth, which adds the folklore and legends associated with mines and caves to the mix of geology and theosophy, he claims that:

> There have been numbers of people who, over the years since the publication of *The Coming Race* in 1871, have believed it to be literally true – the description of an actual race of people living below the surface of the world. But of these believers, few were more passionate in their conviction than Adolf Hitler. (101)

The claim that Hitler, who we have already seen associated with the Koreshan version of the hollow earth, was also a believer in the power of vril is one that seems to have first been suggested by Willy Ley in a 1947 article for *Astounding Science Fiction*. This claim was brought to the attention of a more mainstream audience in *The Morning of the Magicians* (Pauwels and Bergier), in which the authors claimed to have learned that a secret community was formed in Germany prior to the Second World War whose followers took the Bulwer-Lytton novel as fact. This Berlin group, it was said, believed that a pure Aryan race dwelled at the earth's core and possessed a secret knowledge that gave them great power. This group is said to have called itself 'The Luminous Lodge' or 'The Vril Society.' From there the claims of Nazi associations with hollow earth lore grow exponentially. It is said that Hitler was a believer in the theosophical lands of Hyberborea, Lemuria, and Atlantis and sent expeditions in search of them. He is also said to have believed in the subterranean land of Agharta, ruled over by the mystic city of Shambala, and to have thought that a passageway to the center of the earth was located in Tibet as well as at the poles. Seeking vril power as his own, Hitler is presented by some as immersed in the search for the lost worlds and hollow earth of esoteric traditions.

Rob Arndt, for example, claims that the Vril Society, and its sister Thule Society, which believed that the origins of the Aryan race were in the northern land of Hyperborea or Thule, initiated a program in the 1920s to create flying saucer technology based on that used by the underground races. He claims that the first device built using vril-based technology was completed in 1922. Not an aircraft, it was what he calls an inter-dimensional flight machine. (http://laesieworks.com)

Terry Melanson goes into even greater detail in describing the history of vril in Nazi Germany in his 'The Vril Society, the Luminous Lodge and the Realization of the Great Work' (www.conspiracyarchive.com). Melanson claims that the Vril Society

> combined the political ideals of the Order of the Illuminati with Hindu mysticism, Theosophy and the Cabbala. It was the first German nationalist group to use the symbol of the swastika as an emblem linking Eastern and Western occultism. The Vril Society presented the idea of a subterranean matriarchal, socialist utopia ruled by superior beings who had mastered the mysterious energy called the Vril Force.

Melanson reports that the Nazi flying saucer program was made possible by a secret arrangement between the Vril Society and the Vril-ya themselves to share technology. He writes:

> As early as 1936 Hitler was sending teams of 'Spelunkers' into caves and mines all over Europe searching for the Vril-ya. The Nazi's had also explored Antarctica extensively during the years 1937–38. In search of the fabled hole of the South Pole they apparently had success, like Admiral Byrd, in discovering these entrances. It was here that some say they made contact with the 'Unknown Superman' who lived in the fabled 'Rainbow City'.

Though discussion of the claim seems to have fallen off in recent years, Mattern Friedrich, in the 1970s, claimed that Hitler had escaped alive following the end of the Second World War. He supposedly made his way to Argentina and then to the hollow earth through the southern polar opening. Many, if not all, present-day UFOs are actually craft controlled by the subterranean Nazis.

Of course, the power used by Nazis for their saucers is also available to you – for the small price of $75. Yes, you heard me right. The Vril Generator can be yours. According to one website:

The Immortals of the Subterranean Kingdoms of Light, Agartha, are NOW announcing its reality to Surface Dwellers

The VRIL Generator, is a blessed gift from 'the high astral technology of Agartha', by the surface and inner earth residents of the Agarthian lineage. It is a synchronizer of the high frequency of energy-light, once activated in the inner space of self-originated consciousness, this energy is revealed through the act of practice, connecting simultaneously with the inner door, opening one's own inner and multidimensional reality to the external dimensional doors, which opens the space-temporal gates between the real world and the virtual one. Connecting One to the VRIL power, the hidden energy and awareness of Immortality. (http://onelight.com)

FINDING THE HOLLOW EARTH

In addition to employing vril power, it is also possible for us to make contact with, and perhaps meet, the people from the center of the earth. Diane Robbins, in her book *Messages from the Hollow Earth*, describes how she came to channel messages from the subterranean world. Robbins claims that she was first led to make contact with the subterranean races after learning about a woman named Sharula, who was born in Telos, a subterranean city located under Mount Shasta. Sharula, now a resident of Santa Fe, New Mexico, published a newsletter detailing life in Telos and describing the High Priest Adama. Soon after learning of Telos and Adama, Robbins was contacted telepathically by the high priest. Adama then connected Robbins with Mikos, a dweller not in the subterranean caverns under the earth's surface but in the earth's hollow interior. Mikos lived in the city of Catharia, located in the hollow earth under the Aegean Sea. As Robbins puts it, 'I was talking to Adama one day, when Mikos got on the line. It was a 3-way conference call' (11).

Robbins learned much from Mikos about life in the hollow earth. The residents of the hollow earth travel in electromagnetic vehicles that levitate a few inches above the ground. The capital city is known as Shamballa, located at the very center of the planet. The great power source for these people is a form of free energy – the power of the universe itself (vril?). Crystals combined with electromagnetism offer a way of channeling that power.

❬ THE GREAT POWER SOURCE FOR THESE PEOPLE IS A FORM OF FREE ENERGY – THE POWER OF THE UNIVERSE ITSELF ❭

In addition to the openings at the North and South Poles, the hollow earth can be reached from many caverns around the world and many people have traveled back and forth between the two realms. The great inventor Nikola Tesla, for example, now lives inside the planet. Once inside the hollow earth, visitors find that they are able to travel from there to anywhere in the universe, through time and space portals. There are also spaceports inside the hollow earth that house ships bound for the stars. Also inside the hollow earth is the great library of Porthologos, which contains all the knowledge of the universe. This library is only a physical manifestation of the spiritual library that dwells within each of us, however. The living library of knowledge is located within and accessible to each of us, no matter where we are. She also learned that the earth itself is really a giant crystal and that everything in the hollow earth is constructed of crystals.

According to Mikos, important events will soon come to pass. The earth is shifting into higher dimensions and the great day of promise is soon to dawn upon our planet. When this happens disease and war will be no more and the races of the underworld will unite in peace with the people of the surface. Mikos describes the great event: 'This is the day we will open up the tunnel exits and come to you, with our brightly colored robes and sparkling sandals, bearing gifts of immense riches and the necesarry devices that will return your planet to its pristine state once again' (146). Anyone who wishes to be a contactee for the emerging races has only to make their wishes known through prayer, and their names will be added to the list. Points of emergence will coincide with the location of volunteer contactees, so no one need feel left out. When that day comes, the subterranean races will emerge from the caverns and caves and meet with the contactees, bringing joy and peace to all the world.

Of course, for those of us who are a bit more literal minded, Robbins' means of contacting the hollow earth might seem a little far-fetched. There is bound to be a way to reach the earth's interior without resorting to spiritualistic and new age methods like channeling. That might be a fine technique for the vril–Agharta–Shambala wing of hollow earth theory, but not for the Symmes–Reed–Gardner wing. Symmes, you may recall, had proposed a government-funded expedition to find the polar opening. If this is more to your liking, then have no fear, such a voyage may well be in your future. One such excursion has been planned by Advanced Planetary Explorations (APEX). Plans for the trip, originally scheduled for June 26, 2007, came complete with a travel itinerary that included such amazing details as:

Days 9–11	Start the search for the North Polar Opening to the Inner Continent.
Days 12–14	Once found, travel up Hiddekel River to the City of Jehu.*
Days 15–16	Take a monorail trip to City of Eden to visit Palace of the King of the Inner World.
Days 17–18	Return trip back to City of Jehu on the monorail.

*Please note that if we are unable to find the Polar opening, we will be returning via the New Siberian Islands to visit skeleton remains of exotic animals thought to originate from the Inner Earth. (www.phoenixsciencefoundation.org)

The expedition had room for:

- 33 Scientists and Engineers conducting experiments and measurements.
- 15 seats for filmmakers, photographers, and historians.
- 5 seats for IT experts and satellite uplink experts.
- 23 seats for Exopolitical trainers and ambassadors.
- 24 seats for previous team members and truth seekers.

The expedition was to be led by Dr. Brooks A. Agnew Ph.D., who describes himself as 'having been privileged with the highest callings there are: that of teacher, scientist, engineer, talk-show host of X squared Radio, but most of all, I am an Earth Explorer' (from the video *NPIEE: The Greatest Geological Expedition in History*).

*   *   *

Some people make psychic contact with the civilization at the earth's core. Some join expeditions that sail to the North Pole in search of the great entranceway to the subterranean world.

Not me. I pursue weird, unconfirmed stories about caves in Arkansas that lead to subterranean realms inhabited by peaceful Blues and violent Sasquatch. I crawl through mud, swim through freezing water, swing from ropes, crawl between rocks like a worm, and eat dirty, wet trail mix on a seven-hour journey into and out of Blowing Cave.

I do not discover any strange blue people.

I do not meet any high-tech Sasquatch.

I do not encounter any giant lizards.

I do tire myself so completely that I am almost unable to get out of bed the next day – my muscles ache – my head pounds. Every now and again for the next few days I find myself suddenly experiencing feelings of claustrophobia. I shiver for no apparent reason.

Slowly it begins to dawn on me. The truth was right before my eyes. I did meet the Arkansas Cave People! I did meet a strange race of troglodytes with customs beyond my understanding. Now I realize the truth. The Arkansas Cave People are not blue, nor are they inordinately hairy. They are a group of young professionals from Little Rock. To look at them you might never imagine how weird they really are. They explore caves for FUN!

THE SHAVER MYSTERY

The story of the subterranean passage at Blowing Cave is a story that originated in the pages of *Shavertron* magazine, published in an off-again/on-again manner by Richard Toronto. *Shavertron* bills itself as 'Your Only Source of Post-Deluge Shaverania,' and, I must say, it lives

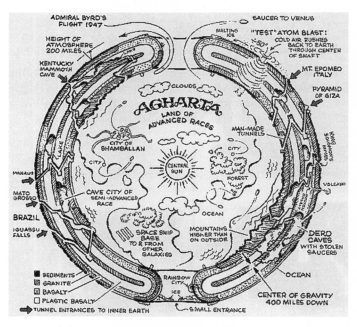

20. A map of the Hollow Earth, with great detail, including the path of Admiral Byrd's flight to the inner world, a saucer on its way to Venus, and the caverns of the hideous Dero.

up to that name. *Shavertron* is devoted to the life, theories, and art of Richard Shaver, a not-so-well-known writer of pulp science fiction stories in the 1940s. He is much more than that, however. His stories were sold as true by Shaver and his *Amazing Stories* publisher Ray Palmer and proved to have a lasting impact on the world of esoterica. Indeed, the Shaver Mystery, as the events surrounding his stories came to be called, is credited by some as providing the impetus for the flying saucer craze of the late 1940s.

Shaver was introduced to readers of Amazing Stories in 1943 when he sent a letter to editor Ray Palmer claiming to have discovered an ancient alphabet that he called 'Mantong.' Palmer published the letter as well as Shaver's story 'I Remember Lemuria.' The Lemuria story was typical of the genre and was supposed to be an account of Shaver's past-life experience in the ancient land. As it turned out, Lemuria would come to provide a back-story for one of the weirdest and most bizarre tales of all time. Shaver himself described the origins of his subterranea: 'Once I was just a person, as you probably are. Then one hot night I had a dream that changed my entire life. Let me tell you the dream and what it did to me' (Swartz, 20). The dream was of a beautiful woman, named Nydia:

> A woman's white face floated before me, the dark red lips hanging like moist fruit, her teeth gleaming. A husky whisper importuned, and I followed . . . Nydia . . .
>
> Every past image of the desired bodies of women, all women I had ever known stretched out their hands to me, and writhed gleaming limbs in a slow dance of longing, holding their red tipped breasts tightly with their spread hands as though to keep the growing buds from blooming suddenly into some wild flower of culmination – and all their faces were . . . Nydia! . . .who was this Nydia I had met in my dreams? (21)

In what has to be one of the weirder passages in modern literature, Shaver describes awaking in the morning following this dream to find that he had overturned the urinal which he kept under his bed (a device that Shaver calls a 'thundermug'). The urine had spread across the brown linoleum on his floor and dried in the perfect likeness of the face of Nydia!

Following this dream, Shaver began to hear voices during his work as a welder on an assembly line. He used two different welding guns in turn, each for a different type of weld that he had to make. He would use one gun, push it out of the way, and then duck as the next one swung into position. One of the guns began to have a very bizarre effect on Shaver, allowing him to hear voices and to know what was in the minds of his co-workers:

> I began to notice something very strange about one of the guns. Whenever I held it, I heard voices, far-off voices of endless complexity. When I changed to the other gun of the pair, I heard nothing. Then I grabbed the gun again as it came around, and right away I knew what was in Bill's lunch box; which girl Bumpy was going to take out that night; the gift Hank's mother was planning to give his wife. It was a dress, and quite a dress too. (22)

The strange effect of the welding gun would not remain so benign, however. Soon Shaver began hearing the far-off voices with added clarity. They were not always pleasant:

> I would hear a mean kind of voice say something like 'Put her on the target.' Then I would hear a woman's screams, louder and louder and more and more agony in the screams – and at last a gurgle, a death rattle . . . I would hear a woman cursing and the lash of a whip – and felt a pleasure in the scream of the person getting the lash. (23)

Soon, Shaver began to suspect that he was being specifically targeted to experience the painful stimulations. He describes his experiences in 'prison,' probably the mental hospital where Shaver is known to have resided for a while:

> I began to feel pains, and the thought in my head kept laughing at me. Like an imp would laugh.
> The pains got worse. Finally I knew what it was, but I didn't believe it. It was people living in caves under and around the prison, and the people's kids liked to torment us prisoners! Not just pester, but real, genuine, torture, done with some kinds of X-rays. (23)

In time, thankfully, the painful stimulations subsided and were replaced by decidedly more pleasurable forces. Shaver assumed that some authority figure, perhaps a 'lawman,' had discovered the crime and put a stop to it. This was not the full story, however, as Shaver learned when the pleasurable stimulations took on a decidedly more erotic feel.

> I learned that for every pain ray, those secret people had a dozen infinitely pleasant rays. They called it 'stim' and woke me every

THE HOLLOW EARTH * 159

morning with floods of the most delicious sensations on earth. How can I best describe stim? Well, it's like a young girl's kiss augmented by a magic to a million powers. One of the young ray people was a girl who seemed to fall for me. She especially dished out this stim ray to me. There is no value on earth as great as stim love. Her name was Nydia, the same girl of my dream! (24)

On one occasion, Nydia used the stim ray to control Shaver's prison guard. Shaver used the opportunity to escape from the prison into a cavern in the nearby woods: 'The dim light inside the caves I found emanated from long tubes running alongside cavern walls. The cavern was an alien place. Vast sculptured rock-forms of non-human beings supported the high, carved arches of the roof that was a mountain overhead.' Inside the cavern Shaver met his beloved Nydia and was taken deep into the underground world where he quickly learned that true love-making amplified by the stim rays was the most incredible experience of his life. This love-making is described by Shaver in pornographic detail.

❝ INSIDE THE CAVERN SHAVER . . . QUICKLY LEARNED THAT TRUE LOVE-MAKING AMPLIFIED BY THE STIM RAYS WAS THE MOST INCREDIBLE EXPERIENCE OF HIS LIFE ❞

He learned from Nydia and her people that their civilization was descended from the ancient civilizations of Atlantis and Mu, who were themselves descended from Titans, travelers from space who had made the earth their home. All had gone well for the Titans on earth and they created a world of peace and technological wonder. All of this changed when the galaxy moved through a cosmic dust storm. This storm fueled the sun and caused reactions that made sunlight toxic to the Titans. Most of the Titans responded to the threat by abandoning the planet. Some of them sought safety under the protective mantle of the earth's crust, however.

Using already existing caverns, the Titans used mighty earthmoving machines to dig new caves and tunnels that crisscrossed the planet. With them they took their great machines and scientific knowledge in hopes of someday discovering a way to overcome the solar radiation, and once again live on the surface of their beloved planet. (22)

Unfortunately, the radiation continued to have a negative impact upon the Titans. Some groups managed to remain relatively normal.

These became known as Teros. Many of the Titans, however, began to suffer genetic mutations that destroyed both their intelligence and their morality:

> Eventually, this once mighty race was reduced to mutated horrors, retarded in intelligence and social structure. Worse still, these monstrosities still had access to the self-repairing machines of their ancestors. But instead of using them for their intended purposes, the Abandondero used them to satisfy their sick, twisted desires. These are the demons of ancient myth and folklore. (22)

These Abandonderos, or Deros, were responsible for the painful experiences that Shaver had himself suffered, and indeed for many of the ills of present life that the human race suffers. Literally under our feet, a war is waging between the forces of good and the forces of evil – between Teros and Deros.

Later in his life Shaver added some important elements to his theories, including his discovery of 'Rock Books.' In his workshop in Summit, Arkansas, Shaver painstakingly examined rocks in the belief that they contained information from the ancient civilizations of the Titans. He observed: 'Stones are not just stones, they are often fragments left by a very ancient race. Almost every stone on the surface will show traces of some use by man if one looks' (23). He described his experiences with these Rock Books by noting that when they are first examined the images appear to be jumbles of faces and other things lumped together. Shaver developed several techniques by which he was able to draw out the most dominant images for study.

On one occasion Shaver discovered that one of his rocks was an actual working Rock Book. A holographic image, projected from the rock, explained to Shaver that Rock Books were the remains of 'a vast memory storage system from a civilization that has been dead for so long that I do not have the proper information to tell you in how many years that it disappeared' (36). Shaver realized that if he could unlock the secret of the Rock Books the world would benefit tremendously from the knowledge and insight that they contain.

Not everyone was happy with Shaver's progress, however. He awoke on one occasion to find himself in a strange cavern. His arms and legs were chained to the cavern wall. As his eyes adjusted to the darkness, a truly terrible sight appeared before him:

> The room I was in was a hollowed out cavity in rock. Slimy water oozed through fissures and dripped onto the floor. Nearby were a

couple of burning torches hung on the walls, but they did little to penetrate the gloom . . .

Out of the darkness came a sight almost too incredible to believe. Four misshapen creatures lurched forwards, laboring under the weight of a large chair that they carried on their shoulders. And seated in the chair was a sight that is forever burned into my memory. It was the stuff of insanity.

I knew at once that I was in the den of a Dero. The sickening creature sat on his throne like a prince from Hades and regarded me with tiny red eyes that seemed to glow with a hellish fire from deep within his skull.

Its body was enormously fat and bloated, with yellowish-gray, bristly skin covered in seeping sores and lumpy tumors. The face resembled a burn victim with the skin hanging in great, fat folds across the forehead, cheeks and chin. The hands were strong and hairy, ending in long, sharp, and filthy fingernails.

It was obviously a male, as he was completely nude. His penis was almost hidden amongst the rolls of fat that hung down from his belly, but it was clearly grotesque in size and covered with warts like some obscene animal. When he smiled, I could see a mouth full of sharp teeth. Without a doubt he was a carnivore who had not missed many meals in his life.

'See,' it hissed at me through his inhuman teeth, 'am I not as handsome as you imagined?' (62)

Shaver was forced to participate in an orgy with human women, under the influence of the stim ray and warned that he should give the working Rock Book to the Dero at once. Shaver was returned home and was never able to make the Rock Book work again. One hopes that the Dero left him alone.

For those interested in seeing the kinds of terrible/beautiful things that Shaver saw, there is a way that does not involve physical abductions or psychic viewing. Shaver's paintings based on his analysis of the Rock Books are really quite visually stunning. Doug Harvey, in *L.A. Weekly*, described Shaver's paintings, then on display in California, in impressive terms (March 20, 2002). He described how Shaver sliced the rock in cross-sections and stared at them until images appeared and how he would sometimes project the images from the rocks onto canvases and then paint them, creating 'thickly textured interpretations of the complex scenes he had excavated.' Harvey goes on to wonder 'why this fascinating work – which on visual terms alone ranks with the Surrealist paintings of Max Ernst and Jean Dubuffet – hasn't been afforded a more complete retrospective.'

Shaver certainly conjured up some terrifyingly ugly images. He produced some beautiful ones as well.

* * *

By the time we return to the surface world, the world of light and warmth, I am physically (and emotionally) exhausted. I want to simply collapse on the warm earth until the cold and wet has seeped out of my body. Instead, I follow my leaders down into a ravine, running with cold water. Along the way my muddy boots slip and I tumble and smash my arm into a rock. I realize that my boots, clothes and equipment are all covered in mud. The clean water of the creek is to be used to wash some of it off. It is clear, however, as I splash the water onto my clothes that this is a fruitless endeavor. The mud is not coming off. There is simply not enough water and it is not moving with

21. Alex Reid, Shauna MacDonald and Nora Jane Doone
in *The Descent* (2005). This is going to end badly.
Image courtesy of the Kobal Collection.

adequate force, to do any more than just move the mud around. I think that if backpackers were to come upon our group at this moment they would probably mistake us for Sasquatch, or Dero.

Indeed, that is how I feel. I have become a subterranean creature, more wild than tame, more mud than man. One would not think that being covered in mud would be enough to make a grown man cry, but along with the total exhaustion it comes close. I am about to panic. I imagine that my skin is permanently stained, that my body has undergone some sort of chemical change so that I am now one with mother earth, like Swamp Thing, only with cave mud instead of swamp muck, like Sandman, only with mud instead of sand. I have been down to the realm of the Dero, with their evil technology and their insatiable sexual appetites, and I have made it back to the surface dirty, very dirty.

Then I notice one of our group emerging from a small cavern in the side of the hill, the water of the creek flowing all around him. He is clean. He motions for me to come over and he points me into the small cave. I crawl along the creek for just a few feet and then the ceiling of the cave suddenly rises far above my head. The sound of rushing water is deafening. It is a waterfall. I stand under it and am almost knocked back down by the force. I can feel the mud being washed away.

I am clean again.

*   *   *

In the caverns of madness. Is this where it all leads? If we leave our world open for Bigfoot; for Chupacabras and Nessie; for Star People and Forest Friends; for Atlantis, Lemuria, and Mu, is this where we are bound to end up? In the caverns of madness.

It is easy to lose our way. The caverns are dark. The paths are not straight.

What began as science – in the hands of Halley and others – became something weird in the hands of Symmes, Reed, Gardner, and Bernard. In the hands of Shaver, it became absolutely bizarre. Symzonia becomes Agharta becomes Naziland, becomes the caverns of madness. Dero, the evil Dero, ever under our feet. Stim rays and Rock Books beyond believable.

Some seek out this world, this kingdom that lies where hell ought to be. Hitler sought it, or so we are told. Admiral Byrd found the way by plane. Others promise to sail there by boat. Robbins teaches us to prepare for their coming. Yet Shaver gives us the warning. With great pleasure comes great pain. With great promise comes great fear. With great wonders come great terrors. Shaver stands there, imagining the faces of beautiful women in the patterns of dried urine, envisioning torture as sex and sex as torture.

I don't know how many people believe that the earth is hollow, or that subterranean caverns are home to beings both good and evil. I suspect fewer than believe in Bigfoot, fewer than believe in Atlantis. Out of those who believe these things, I don't know how many turn to Shaver rather than Robbins; to detrimental robots, the Abandondero, rather than Shambala. I hope, for their sakes, that their numbers are few.

PART THREE

Alien Brainwaves from Space!

Six

Ancient Wonders

I am traveling to the heart of Mormon country to attend the 2007 *Extraordinary Technology Conference* sponsored by the Tesla Tech Institute. As soon as the plane lands I will pick up a rental car and make my way to the meeting site as quickly as I can. I am hoping to make it in time to see the demonstration of a working flying saucer.

The pilot tells the crew to prepare for landing in Salt Lake City and we are beginning our descent toward the runway. I have done this many times. I know exactly what to expect. Like so many members of contemporary society, air travel has become commonplace for me. It is hard to imagine that only 100 years ago this would have been unthinkable. It is hard to imagine that for centuries, perhaps millennia, people dreamed of sailing among the clouds. All I can think of is that my legs are stiff and that I will be very glad to get off the plane. I anticipate the landing.

The expected bump of terra firma does not come. Instead there is a brief shudder and we begin to gain altitude again. We are not landing.

The pilot informs us that there was another aircraft blocking our runway. Instead of landing, we will circle around and try again. This is more than a little unnerving. I suspect that we all owe the pilot a word of thanks. But we are all right now. I try to sit back and relax.

Our new path carries us over the salt flats of the Great Salt Lake. I see from the air that it is bigger than I imagined – almost a sea, even though the state has been in a drought and the water is very low. The red-and-white landscape must surely look like the surface of Mars or some post-nuclear landscape from a distant dystopian future. I expect to see John Carter bounding across the seabed with the super strength granted him by the low Martian gravity and the thin Martian atmosphere, followed by his faithful calot.

Indeed, Edgar Rice Burroughs, creator of John Carter, lived here briefly before he became the famous author of Tarzan. *Perhaps he drew his inspiration for Mars from this landscape, and from the stories in* The Book of Mormon. *The followers of Joseph Smith came here many years before that. They had been chased across the country and must have decided that this desolate land was a good home for a band of outsiders. It must have looked like they imagined the Promised Land, complete with its own Dead Sea. No one else wanted this land. They would take it.*

Their story began when Joseph Smith was visited by an angel, given an ancient manuscript, and learned the wonders of the ancient past.

VIMAANAS

There gleamed the car with wealth untold
Of precious gems and burnished gold;
Nor could the Wind-God's son withdraw
His rapt gaze from the sight he saw,
By Vis'vakarmá's self proclaimed
The noblest work his hand had framed.
Uplifted in the air it glowed
Bright as the sun's diurnal road.
The eye might scan the wondrous frame
And vainly seek one spot to blame,
So fine was every part and fair
With gems inlaid with lavish care.
No precious stones so rich adorn
The cars wherein the Gods are borne,
Prize of the all-resistless might
That sprang from pain and penance rite,
Obedient to the master's will
It moved o'er wood and towering hill,
A glorious marvel well designed
By Vis'vakarmá's artist mind,
Adorned with every fair device
That decks the cars of Paradise.
Swift moving as the master chose
It flew through air or sank or rose.
(*Ramayana* – Book 5, Canto 8, Tr. Griffith)

India's epic *Ramayana*, like other ancient Indian documents, describes amazing aircraft, at once beautiful and swift. Moving at the will of its driver, these vimaanas are described as chariots of the gods, as playthings of the wealthy, and as war machines bringing death from the air. While mainstream interpretations relegate the vimaanas to the stuff of legend and mythology, there is a long tradition of taking these tales at face value. Many see in the description of the vimaanas evidence that ancient India was at least as technologically advanced as our own civilization.

Adding to the mystery of the vimaanas is a strange text called the *Vaimanika Shastra*, a text with a pedigree that is almost as mysterious as the vimaanas themselves. According to G.R. Josyer, who translated the work into English, the text of the *Vaimanika Shastra* was written by Pandit Subbaraya Sastry in the early years of the twentieth century. The author claimed, however, that he had only channeled a text that had actually been written in ancient India by Marashi Bharadwaja. The content of the channeled document was quite interesting. It was a manual for the design and operation of vimaanas. Josyer, in his introduction, describes Pandit Subbaraya Sastry in glowing terms. He was, according to Josyer, 'a simple, orthodox intellectual Brahmin with spiritual gifts who was esteemed by all who knew him . . . He was a walking lexicon gifted with occult perception' (vii–viii). Despite the tantalizing topic of the manuscript, Josyer indicates that funds were unavailable for its publication until Josyer took it upon himself to make the effort in 1973.

The text indicates that the operation of the vimaanas requires knowledge of 32 secrets, with which the pilot should become well acquainted before attempting to fly the craft. Among other things, 'He must know the structure of the aeroplane, know the means of its take off and ascent to the sky, know how to drive it and how to halt it when necessary, how to maneuver it and make it perform spectacular feats in the sky without crashing' (2–3). In addition to the basics of flight, the pilot must also be prepared with the proper type of clothing and the proper food. Sounding like something one might purchase in a health food store, pilots are encouraged to extract the essence from food and form it into food balls for consumption on long flights. (Now that I think about it, this sounds a little like airline food.)

Details of the construction of vimaanas are offered as well, including details of special mirrors that can protect the craft from harm. Among these potential threats are the pilots of other aircraft:

> When enemy planes with men intent on intercepting and destroying your vimaana attack you with all the means at their disposal, the

viroopya-darpana will frighten them into retreat or render them unconscious and leave you free to destroy or rout them. The darpana, like a magician, will change the appearance of your vimaana into such frightening shapes that the attacker will be dismayed or paralysed. (24)

In addition to information regarding the use of protective mirrors, more basic elements of the crafts' construction are also included:

In order to enable the wings on either side to spread and flap, proper hinges and keys should be provided for, safely fixing them to the sides of the vimaana, and for enabling them to fold and open easily.

The revolving tractor blades in the front should be duly fixed to the heating engine with rods so that they could dispel the wind in front and facilitate the passage of the vimaana. (92)

While many people view this text as little more than a bit of modern science fiction dressed up to look like a document from ancient India, others have taken it very seriously. Among those is David Hatcher Childress, founder of the World Explorers Club and Adventures Unlimited Press, and self-described 'real Indiana Jones.' Childress not only supports the idea of ancient Indian aircraft, he has himself added details to the mystery, as in his *The Anti-Gravity Handbook*. According to Childress, ancient Sanskrit documents have been found in Tibet that contain information concerning the construction of interstellar space-ships. The method of propulsion for these craft, called Astra, was a form of anti-gravity power known as 'laghima.' Laghima is a power source that resides within the human ego. It is the same power that allows the Hindu fakir to levitate. Childress says that it is unclear whether the Astra were ever used to send people to other planets, though there is some evidence that they may have reached the moon. He goes on to claim that the *Ramayana* details a battle fought on the moon against an Atlantean air-ship.

❝ ANCIENT SANSKRIT DOCUMENTS HAVE BEEN FOUND IN TIBET THAT CONTAIN INFORMATION CONCERNING THE CONSTRUCTION OF INTERSTELLAR SPACE-SHIPS ❞

Childress also elaborates on his readings of other ancient Indian documents and offers up clear descriptions of the ancient vimaana. According to Childress, a typical vimaana was a:

> double-deck, circular aircraft with portholes and a dome, much as we would imagine a flying saucer. It flew with the 'speed of the wind' and gave forth a 'melodious sound.' There were at least four different types of Vimaanas; some saucer shaped, others like long cylinders ('cigar shaped airships'). (131)

Some vimaanas were powered by a 'yellowish-white liquid' and others by a compound of mercury. He writes:

> The 'yellowish-white liquid' sounds suspiciously like gasoline, and perhaps Vimaanas had a number of different propulsion sources, including combustion engines and even 'pulse-jet' engines. It is interesting to note, that the Nazis developed the first practical pulse-jet engines for their V-8 rocket 'buzz bombs.' Hitler and the Nazi staff were exceptionally interested in ancient India and Tibet and sent expeditions to both these places yearly, starting in the 30's, in order to gather esoteric evidence that they did so, and perhaps it was from these people that the Nazis gained some of their scientific information! (131)

Childress claims that Soviet scientists discovered hemispherical glass or porcelain cones that contained a drop of mercury inside. These devices, found in the Gobi desert, have been identified as 'age-old instruments used in navigating cosmic vehicles.'

Childress suggests that these craft may have been flown around the earth. If so, the world of the ancient past must have looked a lot more like ours than we can possibly imagine. Did the Raman empire of India have airports in locations around the globe? Childress notes that the Rongorongo writing of Easter Island, still undeciphered, looks amazingly similar to ancient Indian script. Was Easter Island an outpost of the far-flung Rama Empire, its great stone monuments perhaps built by the locals using the anti-gravity technology of the Ramans?

> Was Easter Island an air base for the Rama Empire's Vimaana route? (At the Mohenjo-Daro Vimana-drome, as the passenger walks down the concourse, he hears the sweet, melodic sound of the announcer over the loudspeaker, 'Rama Airways flight number seven for

Bali, Easter Island, Nazca, and Atlantis is now ready for boarding.
Passengers please proceed to gate number . . .') (132)

In *Lost Cities of Ancient Lemuria and the Pacific*, Childress claims that
the oral tradition of the islanders describes the huge statues as having
moved from the place where they were quarried to their current
position along the shore by walking – specifically by walking in a
clockwise spiral around the island. He writes:

> There was certainly something more than statues on sleds, or even
> walking them on ropes. Even mental power would not have to
> conform to such a strange law as would have the statues walk in a
> clockwise spiral around the island.
>   . . . The only explanation that made even reasonable sense to me
> was the the statues were somehow moved using the natural earth
> energies and possibly the magnetic anomaly in the crater of Ranoi
> Aroi. Therefore, in this theory, the ancients used some sort of natural
> 'anti-gravity.' (320)

Erich von Däniken, in *Chariots of the Gods*, suggests that evidence
of the aeronautical history of the island can be found in the name
given to the place by its original inhabitants, 'The Land of the Bird
Men.' He writes: 'An orally transmitted legend tells us that flying men
landed and lighted fires in ancient times. The legend is confirmed by
sculptures of flying creatures with big, staring eyes' (111).

The technology of the vimaanas was reportedly brought into
the modern world in 1895, some eight years prior to the Wright
Brothers' famous flight. According to one website, Sanskrit scholar
Shikvar Bapuji Talpade built and flew an unmanned aircraft called
Marutsakthi (Power of the Air) in that year (augustmystery.com).
Based on Vedic technology, the aircraft is reported to have flown to
a height of 1,500 feet before a large group of observers. The aircraft
made use of the mercury vortex engine described in the *Vaimanika
Shastra*. This website goes on to report that NASA is currently
attempting to develop spacecraft powered by such an engine:

> The aircraft engines being developed for future use by NASA by some
> strange coincidence also uses mercury bombardment units powered
> by Solar cells! Interestingly, the impulse is generated in seven stages.
> The mercury propellant is first vapourised, fed into the thruster
> discharge chamber, ionized, converted into plasma by a combination
> with electrons, broke down electrically, and then accelerated through

small openings in a screen to pass out of the engine at velocities between 1200 to 3000 kilometres per minute! But so far NASA has been able to produce on an experimental basis only a one pound of thrust by its scientists a power derivation virtually useless. But 108 years ago Talpade was able to use his knowledge of *Vaimanika Shastra* to produce sufficient thrust to lift his aircraft 1500 feet into the air!

Unfortunately the British government did not approve of Talpade's experiments in flight, perhaps because they did not like to be shown up by Indian scientists. After a warning from the British government, the Maharaja of Broda refused funding for Talpade's experiments. His work was forced to come to an abrupt end. Penniless, Talpade was forced to sell the remains of the craft to 'foreign parties' in order to pay back his debtors. This, complicated by the death of his wife, left Talpade unable to contemplate the continuation of his efforts. 'Talpade passed away in 1916 un-honoured, in his own country. As the world rightly honours the Wright Brothers for their achievements, we should think of Talpade, who utilised the ancient knowledge of Sanskrit texts, to fly an aircraft, eight years before his foreign counterparts.'

According to some ancient technologists, India was not the only ancient civilization to achieve flight. Lumir G. Janku, for example, argues that there is plentiful archaeological evidence to support the theory that flight was achieved by the ancient Egyptian and Central and South American civilizations. He cites the discovery in an Egyptian tomb in 1898 of a wooden object first labeled 'wooden bird model.' However, closer analysis revealed elements that seemed to suggest that the original classification was incorrect.

According to Janku, the model:

has the exact proportions of a very advanced form of 'pusher-glider' that is still having 'some bugs ironed out'. This type of glider will stay in the air almost by itself and even a very small engine will keep it going at low speeds, as low as 45 to 65 mph., while it can carry an enormous payload. This ability is dependent on the curious shape of wings and their proportions. The tipping of wings downward, a reverse dihedral wing as it is called, is the feature behind this capability. A similar type of curving wings are implemented on the Concorde airplane, giving the plane a maximum lift without detracting from its speed. In that context, it seems rather incredible that someone, more than 2,000 years ago, for any reason, devised a model of a flying device with such advanced features, requiring quite extensive knowledge of aerodynamics. (http://ancientx.com)

Other ancient civilizations have left behind evidence that air travel may not be exclusive to modern civilizations. Gold objects discovered in Central and South America dating back more than 1,000 years also seem to be models of aircraft. Once again, as in Egypt, the objects were originally thought to be models of animals, in this case birds or fish, and, once again, closer examination reveals them to be much more. Janku argues that, 'When all the features are taken into an account, the object does not look like a representation of any known animal at all, but does look astonishingly like an airplane.'

Janku also finds evidence for ancient aircraft in the pages of the Hebrew Bible. The Book of Ezekiel, for example, describes what might be seen as an encounter between a man of a non-technological culture with an aircraft from a more advanced culture. He writes, 'There is no shortage of descriptions of flying machines in ancient sources. If we try to extract the core of myths of different provenience and remove the embellishments, we discover to our surprise that flying in ancient times seems to be the rule, not the exception.'

Of course, one would expect that if ancient civilizations could fly through the air then they, like their modern counterparts, would discover ways to use such aircraft for the purposes of warfare. Examining passages from the Indian *Mahabharata*, Desmond Leslie and George Adamski believe that evidence of the use of aircraft to deliver nuclear weapons can be found. He quotes an interesting passage from the ancient document:

> A blazing missle that possessed the radiance of smokeless fire was discharged. A thick gloom suddenly encompassed the hosts. All points of the compass were suddenly encompassed in darkness. Evil-bearing winds began to blow. Clouds roared into the higher air, showering blood. The very elements seemed confused. The sun appeared to spin round. The world, *scorched by the heat of that weapon*, seemed to be in a fever . . . The very water being heated, the creatures who live in the water seemed to burn . . . Huge elephants, *burned by that weapon*, fell all around. (95–6)

Leslie takes this to be the description of the use by the ancients of some form of cataclysmic weapon. He finds other indications that the weapon in question was radioactive, particularly a passage that describes the soldiers' response to the blast. According to Leslie, the soldiers who survived the blast rushed quickly to the nearest water. There, they stripped off their armor and washed it and themselves:

22. Does this Easter Island statue depict a man dying of radiation sickness?

Now why should they stop to have a bath and wash their armour in water ... at a moment like this, unless they were frightened of being contaminated by something – unless some peculiar quality had been imparted to their armour by the blast that they knew would be fatal unless swiftly counteracted? (97)

In other words, Leslie thinks that the ancient soldiers were trying to protect themselves from the dangerous radiation unleashed by an atomic bomb.

Childress believes that there is evidence on Easter Island for the use of nuclear energy by the Rama Empire. For example, the *kava kava* statues found on the island depict emaciated figures. According to legend, these figures were inspired by gaunt bodies lying upon the island shore, as seen in the vision of an ancient king. Childress (1988) describes his encounter with the strange figures:

Looking at the shrunken, almost mutated figure, I could not help but think that they bore a striking resemblance to someone dying of radiation sickness ... Was it possible that this was some sort of bizarre memory of radiation victims from that global war of pre-history? If the king was psychic, maybe he somehow saw the 'Spirit' (after all, a kava kava is a spirit, according to the islanders) of one of these victims. Suddenly, I wondered again about all those vitrified masses of rock around the island, particularly the one near Anakena Beach that I had seen which was surrounded by a massive wall. Was a major battle in the supposed war between Atlantis and Rama fought on Easter Island? (323)

*   *   *

I am running late. My rental car does not travel at the speed of a vimaana, especially not the kind I can afford on my budget.

I have already missed a demonstration of a working model of a flying saucer, but I am told that I didn't miss much. The model had been damaged in transport and did not actually fly. Perhaps it had been cargo on my own flight, parts shaken loose in our near-collision with another plane.

I do make it in time to hear a lecture by Dan Davidson, the author of Shape Power, *a book that investigates the power of shapes to focus and alter energies. I settle into my seat in a large auditorium on the campus of a junior college. I am a little surprised at the number of people here today. I have to admit, most of them look like eccentric inventors or mad scientists. There are no lab coats, but there are lots of bad haircuts, ill-fitting polyester* shirts, horn-rimmed glasses, and pocket protectors. The group seems to represent all ages, though they are almost entirely men. Out of 100 or so people in the audience, I see three women. Davidson himself is dressed better than the average audience member and looks healthy and fit. He is doing a good job of holding the attention of the crowd. Many of them are furiously taking notes.

According to Davidson, 'Shape converts the universal ether, the space energy, into other forces such as electrostatic, gravitic, and magnetism.' Today, Davidson is discussing the Joe Cell, something that I know a little about. The Joe Cell, named after a mechanic called 'Joe X' who helped to perfect it, consists of a cylinder composed of concentric stainless-steel pipes. When filled with simple water and attached to an automobile engine it can reportedly increase fuel efficiency by as much as 70 per cent. Of course, mainstream engineers insist that it cannot work, but it has many supporters.

Davidson's thesis is that it is the shape of the Joe Cell that enables it to function. The shape of the cell means that the device serves as a perfect ether collector. Among other things, he argues that his shape theory can explain how the water in the Joe Cell is converted into a combustible fuel without being depleted and how the fuel from the Joe Cell can affect the engine even though it is not directly connected to the engine's airstream.

Davidson also claims that Joe Cells can be used to collect Orbitally Re-arranged Mono-Atomic Elements, or ORMEs. By exposing oneself to ORMEs collected in a Joe Cell, an individual may experience health benefits. Davidson invites members of the audience to look at him as an example. He has been using the Joe Cell in this way for the last several months. Not only has his balding stopped, but new hair has started to grow. The new hair is not coming in gray, either.

Improved gas mileage and a cure for baldness – all thanks to the power of shapes. Of course, Davidson is hardly the first to make such a claim.

PYRAMID POWER

On October 8, 1973, *Time* magazine reported on the growing influence of 'pyramid power' among both celebrities and common folk. It was reported that Gloria Swanson slept with a miniature pyramid under her bed, claiming that it made every cell in her body tingle. Actor James Coburn claimed to meditate regularly inside a pyramid-shaped tent and to place his cat and kittens on a bed placed over several tiny pyramids, theorizing that the kittens might grow up in some unique way because of the power of the pyramids. *Time* also reported that a Texas doctor claimed that microbes placed under a pyramid lived significantly longer than other microbes.

In 1973 Max Toth, president of the Toth Pyramid Company, was selling Pyramid Razor Blade Sharpeners, devices made of cardboard, for $3.50 each. Toth claimed that the device would more than pay for itself by extending the life of razor blades. Marketers of Toth's product claimed that the device could also be used to dehydrate small fish for display purposes. Patrick Flanagan, marketer of his own line of pyramid merchandise, claimed that the device improved his sexual sensitivity. The device was also claimed to dry fruit and vegetables and, conversely, freshen vegetables, restore stale coffee, mellow cheap wine, and improve the taste of cigarettes.

Flanagan himself sold Pyramid Energy Generators, small pyramids attached to a metal base, as well as the Cheops Pyramid Tent, which sold for $25. It was claimed that the tent was 'a good environment for transcendental meditation, biofeedback and yoga, in that it surrounds its inhabitants with energy.' Flanagan claimed to sleep in his tent to improve sexual sensations, but did not officially advertise the product with that claim. He did, however, claim that the pyramid was ideal for food storage. According to Flanagan, the geometric shape of the pyramid acts as a lens that focuses the energy from the earth's magnetic field. The pyramid power can be experienced both within the pyramid and as an emanation from the pyramid.

> ❛ THE TENT WAS A GOOD ENVIRONMENT FOR TRANCENDENTAL MEDITATION, BIOFEEDBACK AND YOGA ❜

Of course, *Time* noted, not everyone will experience these results, as precision is required. Pyramids must be aligned to true magnetic north and their effectiveness may be lessened by placement too close to windows, radiators, fluorescent lights, radios, television sets, and other electronic devices.

Taking a slightly different approach, Christopher Dunn argues that pyramid power offers a convincing explanation for the true purpose of ancient pyramids, an explanation much better than the accepted notion that the pyramids were nothing more than tombs. Indeed, Dunn claims that the ancient pyramids were themselves power plants, used by the ancient Egyptians to power their advanced civilization. Dunn's theory of the pyramids' original purpose began on a trip to Egypt where he was able to examine pieces of granite from the pyramid construction. After examining one piece of granite in particular, Dunn concluded that the pyramid builders must have used 'machinery that followed precise contours in three axes to guide the tool that created it' (220). This artifact pushed Dunn to ponder not just the question of what tools were used to cut the stone, but the question of what was used to guide the cutting tools to achieve such precision. 'These discoveries,' he writes, 'have more implications for understanding the technology used by the ancient pyramid builders than anything heretofore uncovered' (220).

According to Dunn, granite artifacts discovered in Egypt demonstrate to the careful observer that the ancients must have used lathes, milling machines, ultrasonic drilling machines, and high-speed saws. In other words, the Egyptians used power tools to create the pyramids. This, of course, raises the question as to the source of this power. Where are the Egyptian power plants? If they existed why have they not been discovered? Simple analogies with present day power generation may point us in the right direction, says Dunn. Modern-day power plants are some of the largest construction projects completed by our civilization. They are also often located near a water source, for cooling and steam production. With these factors in mind, the pyramids would seem to be our first choice for investigation. Located near the Nile River, they are clearly the largest construction projects devised by the Egyptians. They are the logical candidates for the source of Egyptian power. Dunn writes, 'In light of all the evidence that suggests the existence of a highly advanced society utilizing electricity in prehistory, I began to consider seriously the possibility that the pyramids were the power plants of the ancient Egyptians' (223).

How might the pyramids have originally worked? Dunn notes that the earth's energy comes in various forms: mechanical, thermal, electrical, magnetic, nuclear, and chemical. Each of these forms of energy is a source of sound. The energy of the earth generates different sound waves, each related to the type of energy creating it and the type of material through which it passes. This sound is outside the realm of human hearing, but is clearly present: 'What goes unnoticed as we go about our daily lives is our planet's inaudible fundamental pulse, or rhythm' (224). The pyramids might have worked by vibrating in harmony with the earth's fundamental frequency. According to Dunn, once the pyramid achieved the proper frequency, it 'became a coupled oscillator and could sustain the transfer of energy from the earth with little or no feedback' (225). The smaller pyramids at Giza might have been used to help the Great Pyramid to achieve this resonance, smaller pyramids being used to jump-start the larger.

Stephen S. Mehler finds support for Dunn's theory in what he sees as evidence for a large explosion in the Great Pyramid. He notes that Dunn had suggested that an industrial accident may have ended the pyramids role as a power source for the Egyptians. Mehler notes several features of the pyramid that would seem to support this claim. (www.gizapyramid.com)

First, Mehler notes that the walls of the King's Chamber have separated from the floor and bulge outward, perhaps as the result of a great explosion. Second, a stone box in the chamber is a dark brown color, when it must have originally been the same rosy color as other examples of Aswan granite. This color change may be the result of intense heat. Third, the upper wall of the Grand Gallery shows dark stains that might be the result of heat from an explosion. Finally, there is evidence of explosions in other pyramids. For example, the northwest corner of the Bent Pyramid at Dahshur appears to have been blown away by an explosion. He writes, 'Most of the original casing stones are still intact, yet this one side seems to be blown off.'

Wayne Fenner offers another hypothesis concerning the function of the pyramids. He claims that: 'The Great Pyramid, and perhaps many other similarly huge pyramidions located throughout this planet's land surfaces, are part of a gigantic grid of sophisticated antennae, scientifically and mathematically designed, located and tuned to perform a specific set of communications tasks' (www.gizapyramid. com). The Great Pyramid may be the center of this earth-wide antenna array. Its internal architecture may have been specifically designed to hold equipment used for the tuning of the entire array to 'exact crystal-controlled electromagnetic frequencies.' The pyramids of the earth, tuned by the Great Pyramid of Giza, would function like a giant transmitter, sending signals deep into space.

We know, suggests Fenner, that the pyramid shape has great power. Indeed, a simple cardboard pyramid 'mummifies a half apple supported in its center, at about one-third of its height, while the other half apple placed outside the tetrahedron simply rots.' The power of the pyramid is concentrated by its four sloping sides and the energy is focused in 'the center of the combined perpendiculars of those edges.' We also know that pyramids have been discovered all around the earth, even underwater. These facts combined – the fact of pyramid power and the fact of their placement all over the globe – would seem to indicate that they were meant to function together as a powerful whole.

Of course, Fenner notes, an antenna array may have been only one of the many uses the ancients made of pyramids. It may be that the pyramids were part of a giant, prehistoric plan to genetically modify the life forms of the planet. Perhaps, scientists from another planet arrived on our earth in the ancient past,

> and concluded that this beautiful blue marble was a perfect biological test bed. They genetically modified existing life-forms, mixing some with parts from other creatures from Earth and elsewhere, creating unique and bizarre plants and animals, certain groups of which were sometimes transferred to isolated areas for long-term testing.

The power of the pyramids would not only have been used to help bring about genetic changes, but would also have served as a means to contact the aliens when the experiment was complete. The pyramids were thus built as 'an immensely powerful, planet-wide radio-frequency system to transmit atmospheric, meteorological and biological data extracted from various projects left behind, to their home planet.'

Of course, the function of the pyramids and the need for the use of precision power tools in their construction is not the only mystery that begs investigation. An even bigger mystery, perhaps, is the question of how the ancients managed to move the monumental stones. Mainstream Egyptologists claim that they were moved using human power, levers, and pulleys. Will Hart, writing in *Forbidden History*, is not convinced:

> This is an intractable problem. As long as Egyptologists insist that men lifted up the cyclopean blocks of stone with nothing but brute force and ropes, this problem will need to be overcome. The rest of the construction formula of the Egyptologists is moot until this primary obstacle is dealt with. If they cannot or will not prove that it was accomplished as they claim, then it is time to go beyond challenging the rest of their baseless theories. (210)

The question is not how long did it take to build the pyramid, but rather could the ancients have built it with the tools and technology that mainstream Egyptologists insist upon? The answer is: certainly not.

In 'Levitation in the Great Pyramid?' an unnamed author addresses these questions directly. The author notes that levitation has been suggested as a possible explanation for this amazing feat (gizapyramid.com). Evidence is found for the use of levitation, first of all, in historical sources. According to the author, the tenth-century Arab historian, Masoudi, claimed that the Egyptians had used magic spells to move the stones. This type of magic is also reported in Mayan and Greek legends as well as in the Bible. The tumbling walls of Jericho, it is noted, might even be seen as a kind of levitation in reverse. Levitation may also have provided the means by which the ancient residents of Easter Island transported their huge stone monoliths.

More modern examples of the use of levitation are also available, including the story of John Keely, a resident of nineteenth-century Pennsylvania. Keely claimed to be able to levitate metal balls and other objects as well as to be able to disintegrate granite. The author finds this detail rather interesting:

> Granite contains quartz, which is a crystal, and by causing the quartz to resonate at an extreme rate, it would cause the granite to break up or disintegrate. This rings a bell with some of the research and

23. Coral Castle – was it built using ancient anti-gravity technology?

speculation that the granite in the King's Chamber and the possibility of it producing piezoelectric effects.

It is also reported that Keely produced his effects by using a mixture of copper, gold, platinum, and silver. To cause objects to levitate, Keely would blow a sustained note on his trumpet. Unfortunately, just as Keely was about to market his device an argument with his backers led to the destruction of his notes.

According to the author, a more recent bit of evidence for the power of levitation is the story of Edward Leedskalnin. Leedskalnin built a castle in Florida using blocks of coral weighing up to 30 tons. The castle took 28 years to complete and was built by Leedskalnin without any additional manpower or powered machinery. It is suggested that Leedskalnin may have discovered a way to reverse the effects of gravity. Perhaps Leedskalnin produced a radio signal that would cause the coral to vibrate at its resonant frequency and then employed magnetic fields to 'flip the magnetic poles of the atoms so they were in opposition to the earth's magnetic field.' According to the author, it was undoubtedly through such means as these that the Egyptians were able to create their magnificent structures.

> **❛ PERHAPS LEEDSKALNIN PRODUCED A RADIO SIGNAL THAT WOULD CAUSE THE CORAL TO VIBRATE AT ITS RESONANT FREQUENCY AND THEN . . . FLIP THE MAGNETIC POLES OF THE ATOMS ❜**

Not all pyramids are found in Egypt and South America, it seems. Bosnian-American Semir (Sam) Osmanagic claims that the world's oldest pyramid is found near Sarajevo. He calls it the 'Bosnian Pyramid of the Sun.' According to Osmanagic, the Pyramid of the Sun, and other smaller pyramids in the area, date from around 12,000 B.C. The Pyramid of the Sun is described as larger than the Great Pyramid at Giza. The pyramid, known to locals as Visocica hill, does indeed resemble a pyramid, albeit one covered in vegetation. According to Osmanagic, the evidence for the ancient pyramid runs deeper than first impressions. Granted, the hill does possess a visually symmetric geometry – that is, it does look like a pyramid – but it is also oriented according to the four cardinal points of the compass. In addition, Osmanagic claims to have discovered tunnel systems and chambers within the hill, as well as evidence of stone-and-mortar construction.

Dismissed by skeptics as nothing more than the remains of medieval roadways and buildings, Osmanagic is convinced the ruins are much older.

Though Osmanagic has been careful about giving away too much of his overarching theory of the origin of the Bosnian pyramid, evidence can be found in his online book, *The Origin of the Maya* (www.alternativnahistorija.com). In this book he argues that the Mayans are descendants of ancient Atlanteans who themselves came to our planet from the Pleiadean star system. Relying on the work of Le Plongeon, he believes that the ancient civilizations of Atlantis and Mu represent the origins of human culture. From these original homelands, the Atlanteans spread their civilization around the world, including to Egypt and the Americas. Mayans, the last of the original Atlanteans to survive on the earth, returned to their cosmic home in the ninth century A.D., leaving behind clues to our future and their second coming. The past and future purpose of the Mayans is to:

> *adjust the Earthly frequency and bring it into accordance with the vibrations of our Sun.* Once the Earth begins to vibrate in harmony with the Sun, information will be able to travel in both directions without limitation. And then we will be able to understand why all ancient peoples worshipped the Sun and dedicated their rituals to this. The Sun is the source of all life on this planet and the source of all information and knowledge. (6)

The pyramids that the Atlanteans built were erected on what Osmanagic calls 'energy potent points' and were used to gather and store the cosmic energy that bathes the earth. According to his theory, 'The pyramids erected on these energy potent locations enabled the Maya to be closer to the heavens and to other levels of consciousness' (70).

Assuming that Osmanagic still holds to his Atlantean theory, and he has to my knowledge never repudiated it, we can begin to understand the role he believes the Bosnian pyramid plays. Older than either the Egyptian or the Mayan pyramids, Osmanagic undoubtedly believes that the Pyramid of the Sun is the nearest we have ever come to locating the source of human culture. Since Atlantis and Mu are lost under tide and time, Bosnia represents the nearest we can come to the source of culture, to the Garden of Eden.

This is all made extremely important by Osmanagic's belief that the day of the Mayan/Atlanteans' return is soon approaching, a day when the pyramids of the earth shall again resonate with the vibrations of the earth:

> And with a frequency in harmony, the Earth will, via the Sun, be connected with the center of our Galaxy. These facts become exceptionally important when we realize that we are rapidly approaching December 2012, a date which the Maya have marked as the time of arrival of the Galactic Energy Cluster which will enlighten us. (6)

2012

The Mayan civilization shared with other cultures a penchant for the construction of magnificent pyramids. In recent years, however, the Mayans have gained more attention for their calendars than for their architecture. According to interpreters of the ancient Mayan calendar, December 21, 2012 occupies a very special place in the history of humankind. According to John Major Jenkins, the Maya used two different calendar systems – a short calendar that measured periods of 13, 52, and 194 years, and a long calendar for measuring time on a much grander scale. According to Jenkins, the 'long count' calendar begins on the Mayan date of 0.0.0.0.1. It is thought by many interpreters that this date corresponds to August 11, 3114 B.C. in the Gregorian calendar. The Mayan calendar comes to an end on 13.0.0.0.0, 5,125 years later, or December 21, 2012 (www.levity.com).

According to Jenkins, the important date for the early Mayans was the end date. He claims that they first determined the end date, and then counted backward to establish their own place in the calendar. In order to count back from an end date that falls on the occasion of an equinox, like December 21, the Mayans must have tracked the precession of equinoxes. Jenkins writes:

> The precession of the equinoxes, also known as the Platonic Year, is caused by the slow wobbling of the earth's polar axis. Right now this axis roughly points to Polaris, the 'Pole Star,' but this changes slowly over long periods of time. The earth's wobble causes the position of the seasonal quarters to slowly precess against the background of stars. For example, right now, the winter solstice position is in the constellation of Sagittarius. But 2000 years ago it was in Capricorn. Since then, it has precessed backward almost one full sign.

This slow change would amount to one degree every 72 years, a change that should be noticeable to careful record keepers, even in ancient times. Though many mainstream scholars deny the Mayans' knowledge of the precession of equinoxes, Jenkins thinks that there are many signs that would seem to indicate otherwise.

Jenkins also notes particular motifs in Mayan iconography that help us to understand the importance of December 21, 2012. This date, he argues, represents a close conjunction between the winter solstice sun with the crossing point of the galactic equator and the ecliptic – what the Mayans called the Sacred Tree. So, the Mayans predicted the coming convergence, saw it as a significant event, and thus used it to mark the end of their calendar. The ability of the Mayans to calculate this is something that Jenkins finds astounding:

> It should be noted that because precession is a very slow process, similar astronomical alignments will be evident on the winter solstice dates within perhaps 5 years on either side of 2012. However, the accuracy of the conjunction of 2012 is quite astounding, beyond anything deemed calculable by the ancient Maya, and serves well to represent the perfect mid-point of the process.

Lawrence E. Joseph in *Apocalyse 2012* is likewise amazed by the advanced state of ancient Mayan astronomy. He writes:

> Ancient Mayan astronomy is anything but oojie-boojie. It is a staggering intellectual achievement, equivalent in magnitude to ancient Egyptian geometry or to Greek philosophy. Without telescopes or any other apparatus, Mayan astronomers calculated the length of the lunar month to be 29.53020 days, within 34 seconds of what we know to be its actual length of 29.54059 days. Overall, the two-thousand-year-old Mayan calendar is believed by many to be more accurate than the five-hundred-year-old Gregorian calendar we use today. (12)

Joseph goes on to argue that December 22, 2012, the day after the end, is the day on which the Mayan calendar starts over. December 22, 2012 will be, once again, the Mayan date 0.0.0.0.1. Surely, Joseph and Jenkins both argue, the Mayans knew there was some significance in this date other than that marked by astronomical observations? Surely this day marks a radical transition in human history?

> After centuries of observations, their astronomers came to the conclusion that on the winter solstice of 2012, 12/21/12, or 13.0.0.0.0

by what is known as their Long Count calendar, a new era in human history will commence. This 12/21/12 'stroke of midnight' begins a new age, just as the Earth's completion of its orbit around the Sun brings a new year at the stroke of midnight every January 1. (12)

Joseph expects the year 2012 to be a pivotal event in human history. Perhaps it will be catastrophic, perhaps revelatory – but it will be phenomenal.

In addition to the Mayan calendar's prediction, Joseph sees other evidence of a fast-approaching denouement to human history. Since the 1940s, and especially since 2003, the sun has exhibited an unusual amount of solar activity. Scientists predict it may reach its climax in 2012. These solar storms are related to storms on the earth. As 2012 approaches we can expect to see an increase in the number and severity of weather-related disasters. Furthermore, one of the earth's greatest defenses against solar flares is the magnetic field. The magnetic field has been diminishing, however – perhaps pointing to a catastrophic polar shift.

As if this isn't bad enough, Joseph also notes that Russian scientists believe that our solar system has entered an interstellar energy cloud. This cloud is damaging our sun and the atmosphere of our planet. They predict disasters on the earth in the years 2010 to 2012. Not all of the doomsayers are Russian. Berkeley physicists, who discovered that the dinosaurs were killed by a major meteor impact with the earth, believe that we are now long overdue for another such calamity. Furthermore the Yellowstone Supervolcano, which erupts every 600,000 to 700,000 years, is preparing to blow. This could result in the annihilation of nearly all of the earth's population. Joseph adds evidence from Eastern and Western religions to the depressing list. Both the I Ching and Hindu theology highlight the year 2012 as a critical year in the history of the world. This is also attested to by many indigenous religions, as well as the Bible and the Koran.

Not everyone sees 2012 in such depressing terms, however. Barry and Janae Weinhold, for example, are the founders of the Carolina Institute for Conflict Resolution and Creative Leadership. They look to 2012, not as a dark and ominous date, but as a time of great promise and hope. Janae writes:

> When I think of this shift, I imagine a caterpillar who is unaware he is about to become a monarch butterfly. I see female Monarchs (God) laying us as eggs on the underside of milkweed leaves (Earth).

When we hatched as larvae earthlings, we fed on plant leaves (life) and now we are mature caterpillars. At some point, we get a signal that tells us that change is coming. The caterpillar stops what it has been doing and attaches itself upside down on a branch to spin a pupa or chrysalis that will contain it while it changes into a butterfly. Inside this chrysalis, the substance of caterpillar dissolves into a kind of primordial soup and the coding in its DNA changes it into a whole new being. This chrysalis becomes increasingly transparent during metamorphosis. Then it cracks open and the butterfly emerges. As its wings slowly dry out, it discovers it is no longer confined to crawling. It has wings to lift it into the sky to fly free to live in a whole new reality. Friends, this is what I believe is about to happen to you and me and all of humanity. I've been thinking about this shift for quite a while now and here is what I see. I see a lot of people who are receiving internal signals saying this metamorphosis is already happening. (http://weinholds.org)

I've got to admit, that sounds a lot better.

Weinhold goes on to say that she refers to the process that is now approaching a climax as LOVEvolution. Unconditional love is a subtle or invisible energy that connects the worlds of spirit and matter. When a biological organism is filled with love, it divides and creates more cells to hold the love. This happens at both the basic level of couples having children, and the galactic level, where stars are born. Love transforms DNA and lifts the consciousness of the universe. This coming transformation, 2012, promises to be quite dramatic:

Our DNA is being reprogrammed to trigger our metamorphosis and activate our light body. Time is speeding up because of Earth's changing electromagnetic forces. Some believe that our 24 hour day has been condensed into a 16 hour day. It has become very clear about Earth's place in our solar system and how we are a part of Her. When she is affected by increases in energy coming from Galactic Center, so are we. She is a living being undergoing her own process of growth and transformation. Our development mirrors hers. We and Earth are raising our vibrational frequencies, which is slowly birthing us both into a dimension or reality.

*   *   *

24. The Extra Low Frequency Modulator.

I am having my vibrational frequency raised.

The Extraordinary Technology Conference allows inventors to demonstrate their devices to the public, and I have been enticed to receive a free treatment from a Multi-Wave Oscillator, operated by a tall smiling man dressed in a suit and tie. His device consists of two metal discs, about the size of dinner plates. The discs are painted black with concentric circles of silver. They are placed on either side of me as I sit in a folding chair. I have seen them working earlier and know that the discs will emit little sparks of electricity as they start to spin.

I see him flip a switch and I begin to feel a slight tingle, as if the hair on my arms and neck is standing up. It feels like static electricity on a cold winter morning. This doesn't feel too bad. After the near collision at the *airport and my frantic drive across town to make it here, it is good to relax and let someone else worry about my vibrational frequency.*

The operator moves his hand close to mine and I feel a spark jump between us. The air crackles a bit.

I am told that the device is emitting a wide band of electromagnetic frequencies. Every cell in my body is exposed to these vibrations. What this means is that the different cells of my body, which vibrate at different frequencies, are being exposed to the specific vibrations beneficial to them. Every cell in my body is being vibrated back to its original frequency. Youth is being restored!

But there is more.

The device also serves as an Extra Low Frequency Generator and is actually able to emulate my brainwaves. With practice and meditation it will allow me to artificially alter my brainwaves to reach any desired brain frequency. With it I could achieve a variety of spiritual and psychic effects. The device can even alter my brainwaves so that I may achieve contact with Space Brothers and Sisters. The power of my prayers can be amplified. By stimulating the psychic frequency (7.83 Hz) I can even gain the ability to communicate telepathically with others.

And if I buy it today, I will also get a Violet Ray Wand for the same price. By attaching the Violet Ray Wand to the Multi-Wave Oscillator, the machine's healing energies can be applied to specific body parts or problem areas. It can be used to massage sore feet and muscles and can be applied directly to the breasts and prostate.

2012 suddenly can't come soon enough.

*   *   *

2012 mania seems to be everywhere. Perhaps it is not yet as intense as 'The Millennium' but I suspect that it might yet reach those levels. While there are some, like the Weinholds, who look to this date as a time of glorious change, most seem to approach it with dread. I suppose that it says a lot about the greatness of our own culture that our apocalypses always seem to end badly. When things are bad, people tend to look for the coming judgment day as a time of recompense and salvation. For the oppressed and the suffering, the coming of Jesus is a hope and a promise. It is for the fortunate that his coming is associated with terror – it means that the good things we now have are to be taken away.

It seems that much of the belief in ancient technology is driven by a sense of coming apocalypse – in the bad sense, not the good.

If civilizations in the ancient past could have achieved the same level of greatness that we possess today, or maybe more, and then plummeted from those heights into a pre-modern condition, then the same is possible for us, perhaps even likely. If one thinks of human history as progressive, and believes that, at least in part because of our technology, we are living in the most advanced age of human development, then it is easy to imagine things just getting better and better. Maybe we will, one day, arrive at a glorious utopian age. If, however, history is not progressive, if there have been civilizations as great as ours in the ancient past, then it is possible that we will follow in their footsteps. We too may reach the pinnacle of progress, just to fall backwards again. Our great technology may one day be forgotten, as the human race has to start all over again. In ancient India, the destruction may have been caused by their technology – their use of atomic weapons. In ancient Egypt, the explosion in the Great Pyramid indicates a similar technological disaster.

I suppose that this isn't a bad message for us to hear. If nothing else, it can serve as a reminder that progress is not inevitable, that things can go terribly wrong. It is a reminder that we should respect our technology and neither abuse it nor come to depend on it completely. It is a message of humility. Even our great civilization can pass away, like the great civilizations of the Ramans, the Egyptians, and the Mayans.

If it happened to them, it can happen to us.

SEVEN

Tesla Technology

Even with the power of shapes and multi-wave oscillation, the 2007 Extraordinary Technology Conference is a bit of a downer. On the day that the conference began, the Salt Lake City Weekly ran a cover story that is near to the heart of many of the participants here. The demonstrations continue and the lectures remain on topic, but there is an undercurrent of discussion. I overhear it in the hallways and catch hints of it from the presenters. The newspaper story is entitled 'Fuel Injected Lunatic: Inventor Paul Pantone hoped to save the world. Now, will the world save him?' I read through it for the third time while waiting for another session to begin. Many of the people seated around me are doing the same.

Pantone is the inventor of GEET – Global Environmental Energy Technology. Pantone's GEET device replaces the carburetor on automobiles and is reported to increase gas mileage and reduce harmful emissions. From what I can gather, Pantone has long been an active participant in the Extraordinary Technology group and the Tesla Tech Institute that sponsors it. He pleaded guilty in 2004 to two charges of security fraud. Before his sentencing, however, the judge committed Pantone to the state mental hospital for evaluation. The newspaper article reports that court records show that Pantone 'exhibits grandiose and persecutory delusions, complicated by a personality disorder and a history of substance abuse.' He is thus unable to stand trial for his sentencing.

To complicate matters, Pantone has refused to take prescribed medications, which are believed to be necessary to return him to mental health. Until Pantone agrees to take the medication, he must remain in the mental hospital. If he takes the medication he goes to jail. The article notes that the symptoms exhibited by Pantone that lead mental health professionals to block his release seem to center around a set of core beliefs that Pantone holds and espouses.

Pantone believes that he has invented a technology that will, in essence, allow automobiles to run on water. He also believes that this technology has the potential to save the world from ecological and economic disaster. He fears that the authorities are aggressively trying to discredit him, because of the very real threat that he poses to their wealth, power, and privilege. There is also a confusing bit about Pantone believing that his technological innovation was a gift given to him by an angel, or perhaps only that he thinks of his current wife as an angel given by God to stand beside him in his time of need. It is all very confusing, but apparently in the state of Utah believing that you have been sent by an angel on a mission to save the world is enough to get you committed to a mental institution.

❢ HE THINKS OF HIS CURRENT WIFE AS AN ANGEL GIVEN BY GOD TO STAND BESIDE HIM IN HIS TIME OF NEED ❣

ORGONE ENERGY

Psychoanalyst Wilhelm Reich is an important figure in the story of alternative technology. This student of Freud claimed to have discovered orgone energy – what Charles R. Kelley, in the September 1962 issue of *The Creative Process*, calls 'the creative force in nature.' Reich first discovered orgone within the human body and claimed that his patients received improved mental and physical health by sitting inside devices that he called 'orgone accumulators.' (The U.S. government was less than impressed with Reich's inventions, imprisoning him in 1956 for transporting an orgone accumulator across state lines.) He also came to believe that orgone existed in a free form in the atmosphere and as the creative force of the entire universe. Kelley, in an article that has been both reproduced and plagiarized many times over, argues that it is in its atmospheric phase that the existence of orgone has been most clearly demonstrated. Kelley himself, a former weather forecaster in the Air Force, claims to have put Reich's 'cloudbuster' apparatus to the test. According to Reich, the cloudbuster has the power to dissolve clouds by drawing orgone energy away from them. It could also be used to stimulate clouds to produce rain. According to Kelley, the cloudbuster does everything that Reich claimed it could do. The success of the cloudbuster is only one of the many pieces of evidence for the existence of Reich's orgone energy, however.

Kelley claims that orgone may be observed in the atmosphere under certain conditions:

> All that is required is a small telescope set up near an ocean or lake to look out parallel to the surface of the water between a few inches and a few feet above the water level. The pulsatory movement of atmospheric orgone energy is usually easily observable. Exciting to watch, this phenomenon is completely unknown to orthodox science. It cannot be explained as an effect of wind, for it frequently has a direction cross or opposite to that of the surface wind. (72)

In addition, Kelley claims that further evidence for the existence of orgone energy can be seen by examining the operation of the orgone accumulator. This device, which is built to collect orgone from the atmosphere, is observed to have a higher temperature than the surrounding atmosphere. In addition, humans within an orgone accumulator experience an increase in body temperature:

> Paul and Jean Ritter report the results of 45 separate observations with nine different individuals as subjects. They showed an average rise in temperature of .48 degrees Fahrenheit in consequence of sitting an average of approximately 40 minutes in an accumulator. Keeping subjects ignorant of the purpose of the experiment did not prevent the rise from occurring, but substituting a plain insulated box for the accumulator did. (72)

This increase in body temperature is attributed by Kelley to the accumulation of orgone energy within the human subjects.

Orgone energy possesses several characteristics. It has no mass. It is present everywhere throughout the universe. It is the medium through which electromagnetic and gravitational phenomena operate. It is in constant motion. Orgone contradicts the law of entropy, is the focus for creative energy, is the source of matter, and is responsible for life. It can be accumulated and stored in specialized devices, which also allow the energy to be manipulated and used to alter the weather, heal illnesses, and improve sexual performance and pleasure.

In recent years, believers in orgone energy have claimed that orgone energy can be used to compensate for the negative energy being broadcast into our environment by cell phone towers and satellites. 'Orgone Gifting' is a movement whose purpose is to place orgone devices throughout the environment in order to reverse the effects of this negative energy upon the environment, the weather, and human physical and mental health. These orgone gifts go by many different names, but two terms in recent use are 'Holy Hand Grenades' and 'Tower Busters.' The term 'Holy Hand Grenade' (HHG) places emphasis upon the size of the devices and the ease with which they can be thrown, usually from a moving car, into areas in need of orgone rejuvenation. 'Tower Busters' are placed at or near cell phone towers to reverse their negative impact.

The orgone gifting website, www.cloud-busters.com, offers advice to those who 'discover the miraculous healing and balancing effects of working with orgone energy devices:'

> You know what it has done for you, your family, and friends. You've watched it take "old sourpusses" and turn them into thoughtful angels, conscious of their ugliness. You've watched the ugliness around your home and neighborhood shift into something higher and nobler. You've seen your plants, pets, and water become simply alive after having these things about. And, at some point, you'll want to share this gift with others. (www.cloud-busters.com)

The advice includes a handy list of potential gifting locations.

The first place to place an orgone energy device is on the main water line entering your home. Because orgonized water 'is conductive and spreads, even against the flow of water itself,' this will benefit all of your neighbors and not just your own water supply. 'One pint-sized HHG or larger will be sufficient to handle any volume your residence may use.' You should also put HHGs under your beds, on your refrigerator and water heater, and in your garden. Place them at each corner of your property to 'create a grid zone of energetic energy.' This will help to reduce the influence of microwave transmissions from cell phones as well as from paranormal entities. Place a HHG at the base of trees to 'de-smog' the neighborhood air. You should also place HHGs in your automobile.

When your own house is in order you should place the devices in all of the places that you frequent – your gas station, park, and grocery store as well as at the homes of friends and neighbors. You don't even have to let other people know what you are doing. Just drop them in an inconspicuous location and let them do their magic. When your environment is taken care of then you should plan to 'grid your hometown.' Everywhere you can place a device, do it.

Many have done this with great success and report excellent results from the community – better attitudes, a vibrant and healthy plant and aviary life, and a far better atmosphere all around than was ever there before. What you are doing is freeing the world around you. By this process, you are disabling underground bases, cell towers and their negative effects, black magicians, and satanists who live in your area. You are freeing the many from the dogma of the church and are forcing law enforcement to rethink what they have allowed themselves to become. Parents become better parents, children better adjusted and normalized. The years ahead no doubt will reveal much; as yet, the data is too new and baselines are just now being established. On the whole, all fronts and aspects of life show a much improved increase in vitality and radiance. (www.cloud-busters. com)

Of course, you might also like to target specific areas and should feel free to do so. Some locations that may need special attention include: cemeteries; Masonic, Mormon, and Esoteric Order Temples; police departments; battlefields; hospitals; reservoirs, streams, lakes, and waterways; prisons; microwave and cell phone towers; nuclear power plants; haunted houses; underground tunnels and bases; daycare centers and schools; bars, distilleries, and sex shops; forests and parks; near high-voltage power lines; and along the freeway. And remember, we are told, you don't have to worry about over-gifting because 'There is no such thing as over-gifting.'

THE NIKOLA TESLA STORY

If there is anyone more influential than Reich in the field of alternative technology, it is certainly Nikola Tesla, famed inventor and champion of alternating current. Unlike Reich, Tesla's influence extends beyond the importance of any particular invention or discovery, like Reich's orgone. Tesla's influence is closer to a cult of personality. For example, in 1959 Margaret Storm released *Return of the Dove*, a self-published volume printed in green ink. In this book Storm claimed that Nikola Tesla was really from the planet Venus. According to Storm, Tesla was born on a space-ship en route from Venus to earth, where he was placed in the adoptive care of Jonathan and Martha Kent, I mean Milutin and Djouka Tesla.

Storm claims that the information regarding Tesla's true origins was revealed in 1947 to Arthur H. Matthews of Quebec, an electrical engineer who had been associated with Tesla from boyhood. It was

25. Nikola Tesla reading.

to Matthews that Tesla had entrusted two of his greatest inventions prior to his death – the Tesla interplanetary communications set and the Tesla anti-war device. Tesla also left special instructions to Otis T. Carr of Baltimore, who used this information to develop free-energy devices capable of 'powering anything from a hearing aid to a spaceship.' (73)

Storm offers additional evidence for the extraterrestrial nature of Tesla. For example, Tesla had very long hands, with unusually long thumbs. His hands were extremely sensitive and carried strong etheric currents. He had the ability to see 'the murky gray astral matter which exuded from the hands of the ordinary person, an effluvia of filth so sticky that it will adhere to the etheric structure of another person – even an individual occupying a body of high vibrations. For this reason, Tesla always dreaded shaking hands' (83). I can imagine so.

While on earth, Tesla clearly understood his mission as preparing the planet earth for the space age:

> Tesla had obviously agreed to come to the earth as a volunteer worker to assist in launching the New Age which he knew to be synonymous with the Space Age. It is perfectly apparent that Ascended Master Saint Germain had to bring in people from other planets, people with knowledge of outer space conditions, to handle major aspects of the planned program.

Tesla was designated to work on the third Ray of Love-in-Action, for that is the Ray which supplies our atmosphere with electricity. (77)

When he agreed to come to earth on this mission, Tesla was agreeing to bring light to the earth.

Unfortunately the United States misused the gifts that Tesla offered. The greatest injustice came when Tesla's alternating current technology was used to power an electric chair for the New York State Prison system. This was 'an all-out effort to lower the planetary vibrations by using electricity, the Third Aspect of God, or the Holy Spirit, to electrocute condemned prisoners' (125). Since America allowed Tesla's gift to be misused, it is America that will pay the karmic debt, through a decline in scientific advancement.

Storm's account does not end on this note of despair, however, for she knows that Tesla continues to work for the good of the earth, even after his death. Though Tesla is now deceased, he continues his research in the scientific department of the underground kingdom of Shambala (74). There, he works alongside his 'Twin-Ray, the White Dove, who was his constant companion on earth.' He shares his ideas and discoveries with chosen ones on the surface, like Otis T. Carr. It is through the work of such disciples that Tesla's work goes on toward the dawning of the New Age. In this New Age, 'Tesla Technology' will prevail: 'In the New Age only new age methods will be utilized. Very soon now will come the big planetary housecleaning. Then down will come all the cables, conduits, wires and posts, which the public is now paying to have installed. What fools these mortals be!' (132).

Tesla's technology, through Carr's free-energy device, will revolutionize the world. Instead of purchasing power from the large corporations, which is then delivered to our homes via wires and cables, the new technology consists of nothing more than a small antenna that will be attached to the roof of every building:

> The antenna will pick up a beamed supply of current, just as a radio picks up a broadcasted program. Inside the home or building the electrical service units such as lights, irons, stoves, adding machines, typewriters, and so forth, will be free from wires and plugs for wall outlets. The units will pick up the necessary current beamed from the antenna. (132)

Truly, this will be a golden age.

FROM HEAVEN TO EARTH

In addition to his communication with Matthews and Carr, Robert Leichtman has also claimed communication with Tesla. In *From Heaven to Earth*, Leichtman records channeled messages from 24 famous individuals, including the renowned inventor. As a background to the discussion we are told of some of Tesla's most sensational and secretive inventions, including a death ray, capable of attacking targets up to 250 miles away. Tesla also developed the principle of resonance, the ability of one object to be made to vibrate at the same frequency as another object. As a result of this research, Tesla caused a small, controlled earthquake. If he wanted, he could have split the earth like an apple.

We are told that Tesla was able to achieve such greatness because of his ability to create a mental picture of the device or problem on which he was working:

> As stunning as it may seem, Tesla actually created and *built* prototypes of his inventions in his mind. He generated not only good ideas, but also working models of them. He could switch on these models, let them run for a month or two, then tear them down and inspect them for wear – all in his mind! (15)

26. David Bowie as Nikola Tesla; need I say more?
From Christopher Nolan's *The Prestige* (2006).
Image courtesy of the Kobal Collection.

Among the information that Tesla shares through Leichtman's mediumship is a bit of important information about the relation of the moon to various physical and mental phenomena on earth:

> When it is a full moon . . . it's high tide not only in the oceans but also on the subtle planes. The subtle body of the planet is affected – etherically, astrally, and even mentally. It is high tide, which simply means that matter is stirred up . . . Electromagnetic phenomena are enhanced at the time of the full moon. (22)

He also reveals that he continues to research and create in the after-life: 'I teach in a sort of Electrical College on the inner planes. I instruct many people while they are out of their bodies at night' (48). By teaching students while they sleep, Tesla is returning the favor for his own education:

> Every night for a long period of time I would leave my body while I was asleep and go to classes on the inner planes. I would participate in experiments in actual laboratories, attend lectures which would add to my understanding of the phenomena and principles of electricity, and then come back to earth and wake up. (57)

*   *   *

Between sessions of the conference, crowds tend to gather around one exhibit or another. Now there is a group of 35 or 40 men standing in a semi-circle in front of one of the tables. I hear a few 'oohs' and 'aahs' coming from the crowd. I can't see a thing, so I carefully sneak around behind the table and watch from the wings. A middle-aged man, dressed in blue jeans and a western shirt, is talking a mile a minute.

On the table in front of him are two electric motors. I am not sure that I understand what he is talking about, but the other 'inventors' all seem to. Every now and then one of them will ask a question, which the presenter graciously answers.

From what I am able to tell, his claim is that he has altered the electric motors so that they each provide a power source for the other. Both motors are drawing electricity from an electrical outlet, but, he claims, they are both producing more energy than they are using. By working together each sends its excess power to the other, and vice versa, until a perpetual loop is achieved. You should be able to add more motors to the system and improve efficiency with each addition.

Theoretically, he informs us, he should be able to unplug both motors from the power supply and they would continue to function. There are a few bugs yet to be worked out, however, and that must wait for the future.

If what he says is true, this is pretty amazing. An electrical power source must be used to jump-start the motors, but once it is operating that power source can be disconnected. The motors produce more power than they consume. This would revolutionize the world and solve all of our energy problems.

'Where is the extra energy coming from?' he is asked.

'From the universe itself.'

'Ooh! Aah!'

'I have instructional videos for sale, if anyone is interested.'

Tesla's Amazing Inventions

According to some, Tesla's greatest inventions are being kept from us by powerful government and capitalistic interests. For example, it is believed that Tesla's discovery of free energy has the potential to transform the world and usher in a new age, yet this technology is kept under wraps by the authorities, who are afraid that it will destroy their economic interests. One of the most passionate defenders of free, or 'zero-point,' energy is Tom Bearden. In his article 'The New Tesla Electromagnetics and the Secrets of Free Electrical Energy,' Bearden describes the importance of Tesla's discoveries. According to Bearden, Tesla's electromagnetic (EM) theory drastically changes quantum mechanics, quantum electrodynamics, and relativity theory. Indeed, our present EM theory is really 'just a special case of a much more fundamental electromagnetics discovered by Nikola Tesla, just as Newtonian physics is a special case of the relativistic physics. But in the new electromagnetics case, the differences between the old and the new are far more drastic and profound' (www.totse.com).

Bearden believes that there are free-energy solutions to our energy problems. It is crucial that we begin work on these solutions immediately, because the 'present energy paradigm' has failed. This present paradigm says that we must consume fuel in order to generate energy and that this consumption of fuel must come at some expense to the environment. As long as we continue to look for solutions within that paradigm, we will fail. Based on the work of Tesla and others, Bearden argues that we must simply abandon this paradigm: 'One does not have to consume fuel in order to obtain energy! Spacetime/vacuum itself is the greatest energetic source in the universe, with more energy in a cubic centimeter of space than the Earth consumes in a million years' (www.cheniere.org). Indeed, Bearden himself has already completed the foundational work on this technology and has patented the Motionless Electro-Magnetic Generator. This device is claimed to output between 5 and 20 times more energy than it uses.

In addition to his free energy discoveries, Tesla is also known for the death ray. According to David Hatcher Childress, the ray worked in relation to a charged tower. If any ship or airplane entered the electric field generated by the tower, their electrical system would be destroyed by a beam of high-energy particles. Tesla and Childress also claim that this energy could be transmitted over great distances and with great precision. They write: 'Such a transmitter would be capable of projecting the energy of a nuclear warhead by radio. Any location in the world could be vaporized at the speed of light' (253). Childress believes that Tesla

❝ A TRANSMITTER WOULD BE CAPABLE OF PROJECTING THE ENERGY OF A NUCLEAR WARHEAD BY RADIO. ANY LOCATION . . . COULD BE VAPORIZED AT THE SPEED OF LIGHT ❞

successfully tested the death ray on one occasion, June 30, 1908. On this date a horrific explosion occurred near the Stony Tunguska River in Siberia. The explosion is reported to have flattened half a million acres of pine forest. While some have suggested that this event was caused by a meteor that exploded just prior to impact, Childress believes that the Tunguska Event was the result of a test firing of Tesla's death ray. It is believed that knowledge of Tesla's death ray has been controlled by the U.S. government and that much of the research related to the Strategic Defense Initiative or 'Star Wars' missile defense system was based on Tesla's death ray technology.

The U.S. government has also made use of another of Tesla's amazing inventions. Jeane Manning and Nick Begich argue in the book *Angels Don't Play this HAARP: Advances in Tesla Technology*, that Tesla Technology, used not as he intended but rather for the purposes of evil, is at the heart of the High Frequency Active Auroral Research Program. HAARP, a joint project of the U.S. Army and the U.S. Navy, based in Alaska, uses Tesla Technology to manipulate the ionosphere: 'Put simply, the apparatus is a reversal of a radio telescope – just transmitting instead of receiving. It will "boil the upper atmosphere." After disturbing the ionosphere, the radiations will bounce back onto the earth in the form of long waves which penetrate our bodies, the ground, and the ocean' (8). According to the authors, many 'technological notes' can be played on the HAARP equipment. It can be used to manipulate global weather – for example, the current phenomenon of global warming is attributed to the use of HAARP technology. The device can be used to destroy ecosystems, knock out electronic communications equipment, and, perhaps most diabolical of all, to alter our moods and mental states. HAARP technology could be used to decide elections, sell products, and foment wars.

TESLA IN SPACE!

In addition to death rays, mind- and weather-control technology, and free-energy devices, it is also believed by some that Tesla discovered a method to nullify or reverse the effects of gravity, allowing for the possibility of air and space travel. William Lyne, for example, in his

27. A table-top model of the OTC X-1.

book *Occult Ether Physics: Tesla's Hidden Space Propulsion System and the Conspiracy to Conceal It,* not only claims that Tesla discovered an anti-gravity device but also that this device was used to develop flying saucer technology: 'The flying saucer is a product of Nikola Tesla's life work, his most fundamentally important invention, for which all his other inventions were in pursuit of, though you probably never heard this fact from anyone else' (3).

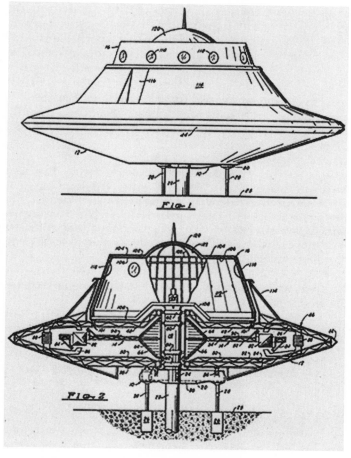

28. Otis T. Carr's patent drawing for his flying saucer amusement park ride.

According to one version of the story, Tesla's anti-gravity saucer technology was perfected in the 1950s by Otis T. Carr. Carr is believed to have followed Tesla's lead to construct several functioning flying saucers. Carr hoped to take one of the craft to the moon. Shortly after a successful test flight, however, Carr's laboratory was ordered closed by the federal government. His equipment, including the saucers, was confiscated. (Another version of the story is that Carr publicized his intention to demonstrate his saucers to the public. Among his development deals was a promise to create a flying saucer amusement ride for the Frontier City Amusement Park in Oklahoma. Carr provided the park with a mock-up of a saucer and promised to demonstrate the real thing at the park. He would then take a larger craft from the amusement park to the moon and return in just a few hours. Neither Carr nor his saucers made their promised appearances. In 1961 Carr was convicted of selling unregistered securities in Oklahoma and was sentenced to time in prison.)

Ralph Ring claims to have been Carr's assistant and co-pilot, and reports that he piloted a 45-foot saucer, the OTC-X1, at the speed of light. He describes the experience:

> Fly is not the right word. It traversed distance. It seemed to take no time. I was with two other engineers when we piloted the 45-foot craft about ten miles. I thought it hadn't moved – I thought it had failed. I was completely astonished when we realized that we had returned with samples of rocks and plants from our destination. It was a dramatic success. It was more like a kind of teleportation. (www.projectcamelot.org)

Furthermore, when the craft was operational, passengers experienced strange sensations:

> . . . the metal turned to Jell-O. You could push your finger right into it. It ceased to be solid. It turned into another form of matter, which was as if it was not entirely here in this reality. That's the only way I can attempt to describe it. It was uncanny, one of the weirdest sensations I've ever felt.

Others believe that saucer technology was well advanced even in Tesla's lifetime, and that it was related to strange transmissions received by Tesla in 1899. Tim Swartz believes that light is shed on the story through information found in Tesla's journals, once thought lost, but recently recovered (at least for a short while). Swartz's *The*

Lost Journals of Nikola Tesla: HAARP – Chemtrails and the Secret of Alternative 4 reports that:

In 1976, four undistinguished boxes of papers were auctioned in the estate of one Michael P. Bornes. Little is known about Bornes except that he had been a bookseller in Manhattan. The auction took place in Newark, NJ, with the boxes and their contents being bought by Dale Alfrey for twenty five dollars . . . He was surprised to find what appeared to be lab documents and personal notes of Nikola Tesla. (11)

Unfortunately, before Alfrey could properly protect the documents, they were taken from his home by the 'Men In Black.' He was, however, able to reconstruct some of the content of the journals, including the information that in 1899 Tesla intercepted messages from extraterrestrial beings. Tesla first believed the beings were preparing humans for conquest and domination.

Assuming at first that the messages were coming from another planet, most likely Mars, Tesla later became convinced that the signals were too strong to have come from anywhere other than nearby space or the moon. He came to believe that the beings were not from Mars or any other planet in our system. He learned from these beings that the temperature of the earth was increasing, and that it would one day be uninhabitable by human beings. Perhaps, he surmised, the messages were from fellow earth people, who had left the earth in prehistorical times, but who still monitored the development of those left behind.

Sean Casteel, in his book *Nikola Tesla: Journey to Mars – Are We Already There?*, offers a slightly different account of Tesla's messages from space. He writes:

Now I have to tell you that this is a story that I can't confirm or deny. I don't have solid pieces of a flying machine or anything like that. It's a really great story, but I can't confirm it. So I'm somewhat loathe to talk too much about it, but I really think that this book could help bring some people forward who may have information about this.

What I have been able to gather is that there was a group of men in the middle to late 1800s, probably after 1850, who were receiving channeled communications from a spirit medium . . .

Allegedly, it started out as a social club . . . They started to get messages from spirits, and eventually communication professing to be an intelligence from the planet Mars. These channeled messages then started feeding technological information that led to the development of early flying machines that this group put together and initially started flying in Northern California. (60)

According to Casteel, in 1899 Tesla himself began to receive messages and decided to participate in the work of the 'Aero Club.' His participation would prove important. Tesla helped the Aero Club to construct a craft that would transport humans to the planet Mars. It was based on 'anti-gravity propulsion' or 'electrogravitics.' In 1903 an attempt was made to send this craft to Mars. Powered by one of Tesla's energy beams, the craft left the earth's atmosphere, but was not heard from again for many years.

Finally, in 1924, Tesla began to receive signals again, but was unable to make any sense of them. He believed that they were signals sent from the space crew to inform earth of their safe arrival on Mars. In 1938 Tesla received direct communication from the Mars project survivors. He learned that the crew had made it to Mars, in part because a ray machine similar to Tesla's had latched onto the craft and drew them in.

❝ THE EXPLORERS FOUND ONLY A DEAD PLANET, THE REMAINS OF THE LOST CIVILIZATION TELLING THE STORY OF THEIR TRAGIC DEMISE ❞

Expecting to find intelligent life, the explorers found only a dead planet, the remains of the lost civilization telling the story of their tragic demise. In time, the Mars mission was able to learn from the technology of the Martians and return to earth: 'These people eventually came back to Earth using devices we now know as flying saucers' (65).

*   *   *

It now appears as if a fight is about to break out between competing vendors.

At one end of the demonstration area someone has set up a very large Tesla Coil. When it was turned on the electrical buzz was audible all over the room. This had been followed by a few loud cracks of electricity. Everyone's attention quickly turned that way.

The Tesla Coil stands about six feet high. From its top protrudes a crown of glass tubes. A man has taken one of the tubes and holds it to within a couple of feet of the coil. As he does, a visible bolt of electricity arcs between the coil and the rod. The air pops and there is a slight smell of ozone.

He then passes the glass rod to another person, who repeats the demonstration.

'This is good for my arthritis,' he says with a grin.

'Oh and for all sorts of other things,' the owner of the device tell us. 'I recommend that you use it at least once a day.'

I decide that I would like to try it, not so much for arthritis as for the snap, crackle, and pop of electricity. I inch closer and prepare to have my turn. As I do, I suddenly feel very funny in my chest. This is weird. I don't have a pacemaker or a bad heart, or anything. But sure enough, there it is. When I step within three feet of the coil, my chest hurts. Tesla's death ray seems to be working, at least on me. I think I will pass, maybe try it again later.

I won't get another chance.

29. A Tesla Coil in action at the
Extraordinary Technology Conference, 2007.

A very angry man walks up and demands that the device be removed. He is the inventor of this particular version of the Tesla Coil and he has a court order to stop anyone else from selling it. He has already been on the phone with his attorney.

The owner of the device looks caught. He turns off the power and begins to shuffle a pile of papers. His accuser approaches him and they began to talk in hushed tones. This is awkward.

Someone speaks up from the crowd and receives a big laugh and a hearty round of applause: 'I thought Tesla invented it,' he says.

The crowd begins to move away. I hear people talking about all the things the coil is good for in addition to arthritis. The list includes time travel.

THE PHILADELPHIA EXPERIMENT

According to conspiracy writer Commander X, Tesla's first experience with time travel occurred in March of 1895:

> A reporter for the New York Herald wrote on March 13 that he came across the inventor in a small café, looking shaken after being hit by 3.5 million volts, 'I am afraid,' said Tesla, 'that you won't find me a pleasant companion tonight. The fact is I was almost killed today. The spark jumped three feet through the air and struck me here on the right shoulder. If my assistant had not turned off the current instantly it might have been the end of me.'
>
> Tesla, on contact with the resonating electromagnetic charge, found himself outside his time-frame reference. He reported that he could see the immediate past, present and future, all at once. But he was paralyzed within the electromagnetic field, unable to help himself. His assistant, by turning off the current, released Tesla before any permanent damage was done. A repeat of this very incident would occur years later during the Philadelphia Experiment. (www. members.tripod.com/uforeview/teslatime/teslatimetravel.html)

Commander X explores these later events in Philadelphia in his publication *The Philadelphia Experiment Chronicles: Exploring the Strange Case of Alfred Bielek and Dr. M.K. Jessup,* a work that relies heavily upon the earlier work of Gray Barker. This strange story begins in the 1950s when Morris K. Jessup, U.F.O. researcher and author of *The Case for the UFO,* reportedly received word that an annotated copy of his book had been sent to the Office of Naval Research. The annotations were written with three different types of ink and apparently by three different hands. The Air Force was so intrigued by the annotated book that 25 copies of the text were produced by the Varo Company of Garland, Texas. A copy of the Varo edition was given to Jessup by representatives of the Air Force. The annotations were, to say the least, strange. Commander X reproduces just a bit of the text from the preface:

> (From Jessup's original work) The subject of UFO's in its present state is like astronomy in that it is purely observational 'science,' not an experimental one; necessarily, therefore, it must be based on observation and not on experiment. Observation, in this case, consists of everything which can be found to have bearing on the subject. *There are thousands of references to it in ancient literature,* but the authors did not know that their references had any bearing, for the subject did not then exist. The writers were recording such things as met their senses solely through an honest effort to report inexplicable observational data.

> (From the annotations) Hoping, in those days, that something would 'come of it.' Nowadays, Science is afraid that 'Something Will Come of it.' It will, too. IN 1956 or 57 the Air force *Will have* Ships LIKE these in appearance & Will 'feel' safe to announce that Human eyes Have seen Saucers from outer Space But to Not be Worried because 'We to have these Ships' Oh! Brother What a farce! ours will be JET propelled not M. propelled. (31)

This whole experience, as you might imagine, left Jessup wondering both what these strange notations could possibly mean and why the Air Force would have any interest in them. It was not long before Jessup had even more to wonder about. In January of 1956, Jessup received the first of several letters from a man identifying himself variously as Carlos Miguel Allende, Carl M. Allen, and Carl Allen. Because of the erratic pattern of punctuation and capitalization, Jessup was convinced that Allende was the author of the earlier annotations. Though more than a little difficult to interpret, these letters appeared to Jessup and others to make several important claims.

First, the following:

> My Dear Dr. Jessup, Your invocation to the Public that they move en Masse upon their Representatives and have thusly enough Pressure placed at the right & sufficient Number of Places where from a Law demanding Research into Dr. Albert Einsteins Unified Field Theory May be enacted (1925–1927) is Not at all Necessary. It May Interest you to know that The Good Doctor Was Not so Much influenced in

his retraction of that Work, by Mathematics, as he most assuredly was by Humantics. (36)

In other words, despite what we are told by mainstream science, Einstein did arrive at a Unified Field Theory but kept that a secret due to his concern for humanity. Indeed, his theory has already been put to the test by Navy researchers:

> YET, THE NAVY FEARS TO USE THE RESULT. The Result was & stands today as Proof that The Unified Field Theory to a certain extent is correct. Beyond that certain extent No Person in his right senses, or having any senses at all, Will evermore *dare* to go . . . The 'result' was completed invisibility of a ship, Destroyer type, *and all* of its crew. While at Sea. (Oct. 1943) The Field Was effective in an oblate spheroidal shape, extending one hundred yards (More or Less, due to Lunar position & Latitude) *out* from each beam of the ship. Any Person Within that sphere became vague in form BUT He too observed those Persons aboard *that* ship as though they were of the same state, yet were walking upon nothing. Any person without that sphere could see Nothing save the clearly *Defined shape of the Ships Hull in the Water*, PROVIDING of course, that this person was just close enough to see, yet barely outside of that field. Half of the officers & the crew of that Ship are at present, Mad as Hatters. A few, are even Yet, confined to certain areas where they may receive trained Scientific aid when they, either, 'Go Blank' or 'Go Blank' & Get Stuck.' Going-Blank IE an after effect of the Man having been within the field too Much, IS Not at all an unpleasant experience to Healthily Curious Sailors. However it is when also, they 'Get Stuck' that they call in 'HELL' INCORPORATED' The Man thusly stricken can Not Move on his own volition unless two or More of those who are within the field go & touch him, quickly, else he 'Freezes.' If a Man Freezes, His position Must be Marked out carefully and then the Field is cut-off . . . It takes only an hour or so Sometimes all Night & Day Long & Worse *It once took 6 months, to get the Man 'Unfrozen'.* (37–8)

So, Allende's story seems to be that in October of 1943 the Navy conducted a test of Einsteins' 'unified field' upon a destroyer. The result was that both ship and crew became invisible. Some of the crew suffered strange effects that continued after the experiment was completed. These effects he calls 'going blank' and 'getting stuck' or 'freezing.' Allende goes on in this and later letters to Jessup to note that the men who freeze experience time differently than the rest of us. These men refer to the experiences as 'Caught in the Flow,' 'Stuck in the Green,' 'Stuck in Molasses,' or 'Going Fast.'

There are very few survivors of this experience. Some went insane. One simply walked through a wall, in plain sight of family and friends, and disappeared forever. Two survivors 'went into the flame.' According to Allende, 'THEY BURNED FOR 18 DAYS' (39). Allende also notes that when the ship disappeared from its dock in Philadelphia it appeared, only a few minutes later, in 'the Norfolk, Newport News, Portsmouth area' (41). It was clearly identified there and then quickly returned to Philadelphia.

Not long after receiving these letters from Allende, Jessup was found dead, officially having taken his own life. Commander X and others, however, believe that he was murdered:

> No matter what anyone says, I have good reason to *know* that Dr. Jessup did NOT take his own life as we have been led to believe, but that he was a 'victim' of 'circumstances.'
>
> Apparently, from classified documents I have seen, Jessup was on the verge of getting someone in the Navy to cooperate with him. This individual was going to cooperate with him. This individual was going to provide all the evidence to prove to the world that the Philadelphia Experiment really did take place. This is the reason Dr. Jessup was SILENCED. If he had lived all this would be out in the open a long time ago. (48)

Following Jessup's death, the story of what came to be called 'The Philadelphia Experiment' continued to develop. It was helped along by a best-selling book by Charles Berlitz and a 1984 motion picture. Things really began to get interesting again, however, in 1989. Commander X reports that it was in that year that: 'a mysterious individual walked up to the microphone at Tim Beckley's annual New Age/UFO fest in Phoenix, Arizona, and shocked the many hundreds in attendance by claiming to have been a survivor of the Philadelphia Experiment' (62).

Al Bielek offered many of the missing details regarding the Philadelphia Experiment, officially called Project Rainbow, including an explanation for the weird effects experienced by the ship's crew. According to Bielek, every human being is born with a set of 'time locks:'

> The soul is locked to that point in the stream of time when conception takes place so everything flows forward at a normal rate of flow at the

time function, mainly the fourth dimension. So you're part of society from day to day. When you wake up at the proper time, you know all the same people, everything's the same, and you haven't slipped into another reality overnight. You're locked into a time span. (67)

Unfortunately the Philadelphia Experiment had ruptured the time locks of the individuals involved. The experiment was carried out by bombarding the ship with energy from four large Tesla Coil devices. Again, Commander X reproduces Bielek's account:

No one in history to that point had ever subjected a human being to such incredibly intense power fields, much less such incredibly intense magnetic fields. Nobody had any idea what would happen. Nobody gave it any consideration except Tesla, who knew something was going to happen . . . It did, of course. They wound up with insane people; they wound up with some people who lost their time locks, walked through nowhere and disappeared forever. They had others who were less fortunate. Four of them who had moved from their original position wound up in the steel deck . . . When the fields collapsed and their time locks were gone, if they had, unfortunately, changed position, drifted and rematerialized in our dimension, our universe (because they were out of it) at a slightly different place or were unfortunate enough to be where the steel decking was, the steel of the deck literally melded with the molecules of the body. Not a very pleasant way to die. (68)

> ❝ SOME PEOPLE WHO LOST THEIR TIME LOCKS, WALKED THROUGH NOWHERE AND DISAPPEARED FOREVER . . . OTHERS WERE LESS FORTUNATE ❞

Over time, the researchers realized that they would need help from an electronic computer to synchronize all of the factors exactly and prevent another disaster. By 1953 they were ready to test again with another ship and another crew. This time it was successful. The program was renamed Project Phoenix and led to the development of much of today's stealth technology. It also led, through a rather circuitous route, to experiments in mind control.

On August 12, 1983, operating from an 'abandoned' military base in Montauk, New Jersey called Fort Hero, Project Phoenix 'locked into the Philadelphia Experiment:'

What happened to the Phoenix Project? It crashed that night, a night of horror in which a well-described monster looking much like an 'abominable' snowman, a Sasquatch, which was described (depending upon how panicked people were who saw it) as 12 to 30 feet high. It went around smashing buildings and people. So the director of the project at that time – it was Jack Pruett – said 'We've got to shut this station down.' (70)

Unfortunately, shutting the station down proved to be impossible; the equipment kept working even when the main power cables were cut.

According to Bielek, this date, August 12, was critically important. For on this date all four of earth's biorhythms peaked at the same time. This had happened on August 12, 1943, 1963, and 1983. Hence the synchronicity between the Philadelphia Experiment of August 12, 1943 and August 12, 1983, and the importance of Al Bielek's role in the whole affair.

Al Bielek claims that he was not born Al Bielek. He was born Edward Cameron on August 4, 1916. He earned a Ph.D. in physics from Harvard and soon thereafter joined the Philadelphia Experiment, then under the direction of Nikola Tesla. It was as Edward Cameron that Bielek participated in the famed event of August 12, 1943. He relates his experience:

> As the experiment wound down, my brother and I walked on deck to see all hell breaking loose. It was horrible. Two sailors were embedded in the deck, two embedded in the bulk head. One sailor has his hand embedded in the steel, which had to be amputated in place. He was the only one who lived and now has an artificial hand. (86)

Cameron/Bielek and his brother tried in vain to shut down the power supply. When this failed they jumped overboard:

> Then we got the surprise of our lives. Instead of landing in the water as you might think, we landed standing up in a military installation in Montauk, New Jersey at Fort Hero – and not in 1943 as one might presume, but 40 years in the future. It was now 1983. (87)

Cameron/Bielek and his brother were told that they had to be returned to their proper time. The Philadelphia Project and the Phoenix Project were locked into a hyperspace bubble. The brothers had to return to 1943 to destroy the experiment:

We were placed in a time tunnel and the juice turned on. When we returned to the *Eldridge* we saw people burning on deck and general mass hysteria. We immediately went in to the control room and axed the equipment. As the machines were winding down – but before the field actually collapsed – (my brother) Duncan again jumped overboard and landed back in 1983. He aged at approximately one year per hour and quickly died of old age ... Duncan was reborn in 1951, sired by our father Alexander Cameron, and his fifth wife. I remained on the ship and returned to 1943. (88)

In time, however, the authorities decided that Cameron/Bielek was a security risk. He was arrested and charged with espionage in 1947. He was then taken to Fort Hero and transported through time back to 1983. He was then sent back to 1927 to be born as Al Bielek. It was not until Bielek saw the motion picture version of the Philadelphia Experiment that the memories of his other life came flooding back. Subsequently Bielek has been instrumental in producing literature related to the Montauk Project – a project reportedly conducted at Montauk, New Jersey and related to the Philadelphia Experiment, mind control, and time travel.

My head hurts.

JOHN TITOR

A standard objection to the reality of time travel is the argument that if time travel is possible then it will surely be perfected at some point in the future. If this is the case, people from the future would certainly want to travel back in time, and not just to the future as Bielek did. In other words, if time travel is possible then we should see evidence of time travelers visiting our time and past periods. Since we don't have any evidence of that, it stands to reason that time travel doesn't work. Of course, one response to this objection is to suggest that time travelers may visit us all the time. Perhaps we simply don't recognize them.

One example of this is John Titor.

In 2000 and 2001 someone calling themselves 'Timetravel_0' and then 'John Titor' made claims on various internet bulletin boards, starting with the Time Travel Institute forum, that he was a time traveler from the year 2036. He made several statements concerning events that were to transpire between 2001 and 2036, including a civil war in the United States in the year 2004 that would result in a fracturing of the nation into five regional governments. Titor's postings caused quite a stir at the time, and even though he has offered no new postings in recent years and his predictions have not been fulfilled, there are some who still believe that Titor is who he says he is.

The postings, reproduced on various websites, began simply enough, with a list of the basic components of a time machine:

02 November 2000 01:00 (about time travel) 1
I saw the posting requesting the basic systems for a gravity distortion system that will allow time travel. Here they are:
1. Magnetic housing units for dual microsingularities.
2. Electron injection manifold to alter mass and gravity of micro-singularities.
3. Cooling and x-ray venting system
4. Gravity sensors (VGL system)
5. Main clocks (4 cesium units)
6. Main computer units (3)

This was followed by a description of what time travel was like for the traveler:

Questions for Timetravel_0 with permission to post.
Pamela: By the way can you tell me what it feels like to time travel? when you are in the process of doing it what does it feel like and what do you see and hear. you made mention that you had to get use to the fields. Do you see a bright flash of light?
Timetravel_0: Interesting first question. The unit has a ramp up time after the destination coordinates are fed into the computers. An audible alarm and a small light start a short countdown at which point you should be secured in a seat. The gravity field generated by the unit overtakes you very quickly. You feel a tug toward the unit similar to rising quickly in an elevator and it continues to rise based on the power setting the unit is working under. At 100% power, the constant pull of gravity can be as high as 2 Gs or more depending on how close you are to the unit. There are no serious side effects but I try to avoid eating before a flight. No bright flash of light is seen. Outside, the vehicle appears to accelerate as the light is bent around it. We have to wear sunglasses or close our eyes as this happens due to a short burst of ultraviolet radiation. Personally I think it looks like your driving under a rainbow. After that, it appears to fade to black and remains totally black until the unit is

turned off. We are advised to keep the windows closed as a great deal of heat builds up outside the car. The gravity field also traps a small air pocket around the car that acts as your only O_2 supply unless you bring compressed air with you. This pocket will only last for a short period and a carbon sensor tells us when it's too dangerous. The C_2O_4 unit is accurate from 50 to 60 years a jump and travels at about 10 years an hour at 100% power. You do hear a slight hum as the unit operates and when the power changes or the unit turns off. There is a great deal of electrical crackling noise from static electricity.

One of the interesting claims that Titor makes is that his time line seems to be slightly different from ours. Football games are won by different teams than are recorded in history books from 2036. This means that his predictions for our future are really only descriptions of his past. These are two different things. The longer he is present in our time, the greater his influence upon the time line and the greater the divergence between the two. The following is an example of the types of claims Titor made:

Pamela: 1. what are some of your memories of 2036?

Timetravel_0: I remember 2036 very clearly. It is difficult to describe 2036 in detail without spending a great deal of time explaining why things are so different. In 2036, I live in central Florida with my family and I'm currently stationed at an Army base in Tampa . A world war in 2015 killed nearly three billion people. The people that survived grew closer together. Life is centered around the family and then the community. I can not imagine living even a few hundred miles away from my parents. There is no large industrial complex creating masses of useless food and recreational items. Food and livestock is grown and sold locally. People spend much more time reading and talking together face to face. Religion is taken seriously and everyone can multiply and divide in their heads.

Titor claimed that he had journeyed from 2036 to the year 2000 as part of a military assignment. He was originally sent to the year 1975 in order to acquire an IBM 5100 computer, which was needed to resolve twenty-first-century computer problems. His grandfather was one of the programmers of the 5100, hence Titor's selection for the mission. He stopped in the year 2000 for his own personal interests and to gather family documents and photographs that had been destroyed in the great civil war. Titor claimed that he had journeyed through time in a device, of which he shared pictures, that was first installed in a 1967 Corvette and then in a 1987 Ford truck.

His last post was in March of 2001, when he announced that he would be returning to the future.

*   *   *

Nine months after my visit to the Extraordinary Technology Conference, Paul Pantone is still in legal limbo. His supporters and defenders are now somewhat better organized. The Friends of GEET – Paul Pantone Defense Project has established a defense fund and has been actively working on Pantone's behalf. Unfortunately, it has not been all that successful. Among various tactics, they convinced the judge in the case to recuse himself from the proceedings following a Christmas ad campaign in which the supporters were asked to send the judge a lump of coal. Neither of these steps met with the approval of Pantone's attorney.

The Friends, however, are concerned that Pantone will never be able to receive a fair hearing. Many seem to believe firmly that he is the victim of a conspiracy to discredit him and his work, thus keeping the world safe for high gas prices. Pantone's attorney, apparently less than convinced of the accuracy of this assessment, has challenged Pantone's supporters to produce a working model of a Pantone engine for his inspection. If he finds it convincing, he promises to film the results for broadcast on television news programs.

Thus far there has been no demonstration.

*   *   *

Wilhelm Reich claimed that his orgone accumulator was based on hard science and that he had evidence that it worked. The U.S. federal government responded by banning the movement of orgone technology or related literature across state lines because they deemed his therapy to be without merit. Equipment was moved and Reich was arrested. Otis T. Carr claimed that his breakthrough in anti-gravity was based on insights from Nikola Tesla. Carr was convicted of securities fraud. Paul Pantone claims to have discovered a technique that will allow automobiles to operate using water as fuel. The State of Utah has declared him incompetent to stand for sentencing for the securities fraud charges to which he pled guilty.

Are these men crooks? Are their discoveries and inventions nothing more than scams, nothing more than tricks to separate gullible investors from their money? Are they kooks? Have they experienced a break with reality that makes them believe their own crackpot theories? Do their financial indiscretions result from a genuine case of misplaced confidence in their own inventions? Are they for real? Do orgone accumulators work? Did Otis T. Carr and Ralph Ring actually fly at the speed of light? Does Paul Pantone have the secret that will change our world for the better? These are not frivolous questions. They are necessary. If we respond too quickly and assume that everyone with a strange idea is a kook or a crook then we surely put a lot of honest, sane people in jails and mental hospitals, or worse. Followers of Jesus claim that it was just this misunderstanding that sent Christ to the cross, after all. Of course, the reverse is also true – if we believe every nutty professor's lame theory, we are bound to be led astray quite frequently.

And then there are the details of Paul Pantone's case. Whether or not he committed the crimes to which he pled guilty is one matter. The question of whether or not believing that his invention really works, that it can really save the earth, that it came to him on the wings of an angel, and that his government is out to get him – the question of whether or not believing all this makes him mentally ill – is quite another. By those standards, there are a whole lot of people in the world – preachers and politicians as well as mad scientists – who should be locked up.

Granted these beliefs are eccentric, they are weird, they are bizarre. But so are the beliefs of a lot of people: Bigfoot hunters, cryptozoologists, friends of the Forest Friends, theosophists, hollow earthers, alternative archaeologists, Christians, Hindus, Muslims, you, me.

I'll admit it, I've been at this so long I'm not sure that I know what is real anymore.

Conclusion

Two memories from a summer day. I am 10 years old.

In one, my grandfather is driving his old truck down Baker Hill. My cousin Barry and I are seated in the back, in the bed of the truck with the spare tire and a few pieces of plywood. Barry gets a look in his eye. I know what it means. We each slip over opposite sides of the truck onto the running boards. We hold on tightly with one hand to the handles on either side of the cab. Facing forward with one foot on the running boards and one hand holding tight to the handles, we throw our free arms and legs out, spread-eagled. We are flying.

In the other memory, we are terrorized by a giant roc, a Thunderbird, a malagor.

Over the course of time I have filed these memories in different boxes, categorized them in different ways. One I treat as a real memory, as the reminiscence of something that really happened in the summer of 1977. The other I have come to regard as a fantasy, as the product of a young boy's overactive imagination.

But to me, many years later, both seem equally real.

*   *   *

William Roe walked out of a clearing in the woods and saw a Sasquatch. Frederick S. Oliver encountered Phylos the Thibetan on the slopes of Mount Shasta. Carlos Allende and Al Bielek witnessed the strange disappearance of a U.S. Navy ship in 1943. In 1977, 10-year-old Marlon Lowe, my contemporary, was playing outdoors in Lawndale, Illinois when he was attacked by a giant black bird. The creature lifted him into the air but was forced to drop his prey when the boy struggled free.

These people have sworn by their memories. They have signed affidavits and gone on the record. They have held onto the veracity of their weird experiences in a way that I have not. I categorized my encounter with a giant bird as a fantasy. Why do others categorize their encounters with Bigfoot, ancient Lemurians, disappearing ships, and, yes, giant birds, as factual reality? Have we responded differently because I am better at logic? Is it because I know something that they don't? In all honesty, I don't remember making the decision to classify it as one thing or another. It was not a choice on my part. As with most things in life, it has simply happened that way. As I have grown older, giant birds have not figured in any other part of my life than that of the fantastic fiction that I still love. Consequently, I suppose, my memory of the roc has been classed as fiction as well.

The difference between my response and the response of others may be related to the fact that my memory comes from childhood rather than adulthood. Perhaps it is harder to shake the reality of such things when they happen to us after we are all grown up. Perhaps if I had seen a roc at age 30, my story would be different, would be harder to relegate to the imagination. This raises the question of whether my response is less reasonable than the others. Is it only because the memory is from my childhood that I can so easily disregard it, so quickly cast it aside with abandoned comic books and forgotten action figures?

But Marlon Lowe, like me, was 10 years old when he saw the giant bird, way back in 1977. Unlike me, he felt himself lifted into the air by its hideous claws and felt the rush of wind from its massive wings and, unlike me, he still believes it to be true, even as an adult. We both share similar memories from a similar time. I have filed mine under fantasy, he has filed his under fact. It is possible that one of us has made a mistake.

*   *   *

People who believe strange things often have good reasons to do so. If someone believes in Bigfoot because they have seen Bigfoot, who am I to call this crazy or to say they have made a mistake? This does not mean that their memories, their experiences, are good reasons for me to believe, however. That is another matter entirely. On the other hand, if I saw a strange, hairy giant looking in my bedroom window, I might believe in

> ❝ PEOPLE WHO BELIEVE STRANGE THINGS OFTEN HAVE GOOD REASON TO DO SO . . . WHO AM I TO CALL THIS CRAZY?❜

Bigfoot as well. Or maybe not. Seeing a giant bird has not made me believe in rocs.

Of course, science demands the kind of verifiability and repeatability that many of the claims and theories talked about in this book are lacking. Science demands the kind of claims that can be accepted from one person to the next because the evidence is open to the public. Despite the work of Meldrum and others, claims about Bigfoot, the hollow earth, and time travel don't meet these standards. It is not that science is predisposed against the veracity of these claims. It is just that the kind of evidence that mainstream science demands is not there. Inasmuch as the discourses in this book are science, inasmuch as they are discourses based on evidence, research, and proof, it is clear that they are of a decidedly weird variety of science. Things do count as evidence here that don't count as evidence for mainstream science, things like unverifiable eyewitness accounts, channeled messages, and ancient manuscripts.

We might even say that the beliefs talked about in this book are more like religious beliefs than scientific theories, despite the scientific clothes that they sometimes are made to wear. We might say that the kinds of evidence counted here are the kinds of evidence that are often appealed to by religious believers. People rely on the testimony of others, on ancient manuscripts and sacred texts, and on visions and epiphanies. Many of the bizarre beliefs talked about in these pages are held in the way that other people hold more traditional religious beliefs. They strike us as bizarre, not because they are qualitatively different from more mainstream religious beliefs, but because they are rare.

*   *   *

For some, the pseudo-science of Bigfoot, Atlantis, Lemuria, the hollow earth, pyramid power, and Tesla Technology is seen as a threat to real science and maybe even to culture itself. Perhaps it is thought that if enough people buy into the unscientific claims of the theosophists and others, then the progress of real science might come to a standstill. Or maybe it is just based on an overriding sense that there is only one way to truth, only one way to arrive at an understanding of the real. This is hardly the position that I would take. I am a big fan of mainstream science, don't get me wrong. I believe that it should be supported by our governments and our universities and that it should be taught to our children. As long as science is kept strong, however, I don't see how Bigfoot and all of the other things talked about in this book can possibly pose much of a threat. I don't see any reason to think that we need to force a choice between one or the other or even get snippy about it in our dialogues and conversations with people with whom we disagree.

I would argue that a culture filled with diverse and idiosyncratic theories and ideas is a good thing. Uniform orthodoxy, whether it be scientific or religious in nature, has proven to be a recipe for cultural stagnation and decline. The printing press kicked off the reformation and ushered in the modern age because it allowed for the dissemination, not of the one universal truth, but of the varied panoply of divergent views that had previously been suppressed. It also helped to usher in the age of science. Suddenly, in a world of competing theories and beliefs, one's ideas had to be demonstrable to others in order to reach consensus, they had to be publicly verifiable and repeatable. It is only against a backdrop of diverse ideas and theories that scientific proof has a place. In a strict orthodoxy there is no place for it to get a foothold. The kind of diverse and idiosyncratic beliefs and theories in this book are, oddly enough, a good thing for mainstream science. It is only in a competitive marketplace of ideas that science is forced to really make its case.

I think it is also the case that a culture of idiosyncratic and diverse theories and beliefs is simply more interesting than a culture of agreement, whether that agreement arises from revelation or rationality. I would not want to live in a culture dominated by religious orthodoxy and I would not want to live in a culture dominated by pure rational thought. Believe in them or not, culture is more colorful thanks to Sasquatch and the gang. I know that *my* life is certainly more colorful.

Of course, another consequence of divergent beliefs and theories is conflict. Modernity not only brought us Protestantism and science, it has proven pretty good at nourishing wide-scale military conflicts as well. Our strong differences of opinions, our radically divergent beliefs, can lead to lots of animosity and bloodshed. It seems to me that it doesn't have to be this way, however. Diverse world views don't have to end in conflict. Much of the animosity could be avoided if we would simply take a more skeptical attitude toward our own beliefs, and admit that we might just be wrong. It also helps if we are humble enough to realize that the world is far more complex than can be accounted for by our own theories and beliefs. Diversity, even the weird and bizarre kinds, helps us in this regard because it reminds us that what counts as evidence differs from context to context. Likewise what is real.

This is a lesson that I have learned on my journeys.

*   *   *

I have tracked Bigfoot through storm-ravaged woodlands. I have talked to eyewitnesses. I have seen the tracks. I have journeyed toward the center of the earth in my quest to locate subterranean civilizations. I have met wise Phylos. I have met the evil Dero lord. I have met beautiful, beautiful Nydia. I have met with the mad scientists. My brainwaves have been altered through multi-wave oscillation. My heart has been touched by Tesla power.

Back in my hotel room, the tracks of the big bull Sasquatch seem to be transformed into nothing more than hoaxes. The dedicated Bigfoot hunters become hoaxers, ironists, or fools. But in the field, kneeling in the Texas mud, the tracks are real. I believe. I feel the tingle of excitement. I feel the rush of adrenaline. I see the honesty on all the faces. The tracks are real.

Down in the depths of the cave, covered in mud and slime, I know that this place holds no secret passages or giant lizards, I know that there are no armed Sasquatch or blue-skinned people. But as I stumble down the hill to wash away the mud, as I slip and fall onto hard stones, it is real. It isn't just mud and exhaustion that are weighing me down, it is the Dero with their diabolical machines.

As I step away from the Tesla coil, it loses its hold over me. Its power ceases to be real. But for that moment, when the arc of electricity leaps like lightning, like a death ray, when my chest grows tight, it is real. Real lightning, real sulfur, real pain, real power.

*   *   *

I open the cover of Burroughs' A Princess of Mars *and read the opening lines to my son: 'I am a very old man; how old I do not know. Possibly I am a hundred, possibly more; but I cannot tell because I have never aged as other men, nor do I remember any childhood. So far as I can recollect I have always been a man.' But I do remember. I have not always been a man; I was once a boy.*

There was a bird. A real bird of prey. Real feathers, red and brown. We ran by the sunflowers that my grandmother always planted at the end of the garden. The sunflowers were real. We ran over the freshly mown grass, wet blades sticking to our feet. The grass was real. We hid under the cottonwood tree, knotty with age. The tree was real.

As I read on as my son's eyes grow wide with wonder . . . as do my own:

I turned my gaze from the landscape to the heavens where the myriad stars formed a gorgeous and fitting canopy for the wonders of the earthly scene. My attention was quickly riveted by a large red star close to the distant horizon. As I gazed upon it I felt a spell of overpowering fascination – it was Mars, the god of war, and for me, the fighting man, it had always held the power of irresistible enchantment. As I gazed at it on that far-gone night it seemed to call across the unthinkable void, to lure me to it, to draw me as the lodestone attracts a particle of iron. My longing was beyond the power of opposition; I closed my eyes, stretched out my arms toward the god of my vocation and felt myself drawn with the suddenness of thought through the trackless immensity of space.

This is not a dream. This is real.

Bibliography

Abelard Productions, Inc. (1990) *The Missing Diary of Admiral Byrd.* (New Brunswick, NJ, Inner Light Publications)

Alper, Frank. (1982) *Exploring Atlantis.* (Thousand Oaks, CA, Quantum Productions)

Andrews, Shirley. (2004) *Lemuria and Mu: Studying the Past to Survive the Future.* (Woodbury, MN, Llewellyn Publications)

—— (2007) *Atlantis: Insights From a Lost Civilization.* (Woodbury, MN, Llewellyn Publications)

Arment, Chad. (2004) *Cryptozoology: Science and Speculation.* (Landisville, PA, Coachwhip Publications)

—— (2006) *The Historical Bigfoot: Early Reports of Wild Men, Hairy Giants, and Wandering Gorillas in North America.* (Landisville, PA, Coachwhip Publications)

—— (ed.). (2006) *Cryptozoology and the Investigation of Lesser-Known Mystery Animals.* (Landisville, PA, Coachwhip Publications)

Bearden, Thomas. (2004) *Energy From the Vacuum: Concepts and Principles.* (Huntsville, AL, Cheniere)

Beckley, Timothy Green. (1992) *Subterranean Worlds Inside Earth.* (New Brusnwick, NJ, Inner Light Publications)

Berlitz, Charles. (1974) *The Bermuda Triangle.* (New York, Avon)

Berlitz, Charles and William Moore. (2004) *The Philadelphia Experiment.* (London, Souvenir Press) Originally published in 1979.

Bernard, Raymond. (1979) *The Hollow Earth.* (New York, Bell) Originally published in 1964.

Binns, Ronald. (1984) *The Loch Ness Mystery Solved.* (New York, Prometheus Books)

Blavatsky, Helena Petrovna. (1877) *Isis Unveiled.* (New York, Theosophical Publishing House)

—— (1888) *The Secret Doctrine.* (London, Theosophical Publishing House)

Bruce, Alexandra. (2001) *The Philadelphia Experiment Murder*. (New York, Sky Books)

Bulwer-Lytton, Sir Edward. (1871) *Vril: The Power of the Coming Race*. (London, Broadview Press, reprinted by Kessinger Publishers 1997).

Campbell, Steuart. (1997) *The Loch Ness Monster: The Evidence*. (New York, Prometheus Books)

Casteel, Sean. (2002) *Nikola Tesla: Journey to Mars – Are We Already There?* (New Brunswick, NJ, Global Communications)

Cayce, Edgar Evans. (1968) *Edgar Cayce on Atlantis*. (New York, Warner Books)

Cerve, Wishar S. [William Spencer Lewis] (1931) *Lemuria: The Lost Continent of the Pacific*. (San Jose, Rosicrucian Press)

Childress, David Hatcher. (1988) *Lost Cities of Ancient Lemuria and the Pacific*. (Kempton, IL, Adventures Unlimited Press)

—— (1999) *The Time Travel Handbook: A Manual of Practical Teleportation and Time Travel*. (Kempton, IL, Adventures Unlimited Press)

—— (2003) *The Anti-Gravity Handbook*. (Kempton, IL, Adventures Unlimited Press)

—— and Richard Shaver. (1999) *Lost Continents and the Hollow Earth*. (Kempton, IL, Adventures Unlimited Press)

Churchward, James. (1926) *The Lost Continent of Mu*. (New York, Washburn)

—— (1931) *The Children of Mu*. (New York, Washburn)

—— (1933) *The Sacred Symbols of Mu*. (New York, Washburn)

Clow, Barbara Hand. (2007) *The Mayan Code: Time Acceleration and Awakening the World Mind*. (Rochester, VT, Bear and Company)

Coleman, Loren and Jerome Clark. (1999) *Cryptozoology A to Z: The Encyclopedia of Loch Monsters, Sasquatch, Chupacabras, and Other Authentic Mysteries of Nature*. (New York, Fireside)

Coleman, Loren and Patrick Huyghe. (2003a) *The Field Guide to Lake Monsters, Sea Serpents, and Other Mystery Denizens of the Deep*. (New York, Penguin)

Coleman, Loren. (2003b) *Bigfoot: The True Story of Apes in America*. (New York, Paraview)

Corrales, Scott. (1997) *Chupacabras and Other Mysteries*. (Murfreesboro, TN, Greenleaf Publications)

Coy, Janice Carter and Mary Green. *Fifty Years with Bigfoot: Tennessee Chronicles of Co-Existence*. (2002, self-published)

Crabtree, J.E. 'Smokey.' (1974) *Smokey and the Fouke Monster: A True Story*. (Fouke, AR, Days Creek Production Corporation)

—— (1999) *Too Close to the Mirror*. (Fouke, AR, Days Creek Production Company)

—— (2004) *The Man Behind the Legend*. (Fouke, AR, Days Creek Production Company)

Daegling, David J. (2004) *Bigfoot Exposed: An Anthropologist Examines America's Enduring Legend*. (New York, AltaMira)

Däniken, Erich von. (1969) *Chariots of the Gods*. (London, Souvenir Press)

Davidson, Dan A. (1997) *Shape Power*. (Sierra Vista, AZ, Rivas Publishing)

de Camp, L. Sprague. (1970) *Lost Continents: The Atlantis Theme in History, Science and Literature*. (New York, Dover)

Donnelly, Ignatius. (1949) *Atlantis: The Antediluvian World, A Modern Revised Edition*. (New York, Gramercy Publishing) Originally published in 1882.

Dunn, Christopher. (1998) *The Giza Power Plant: Technologies of Ancient Egypt*. (Rochester, VT, Bear and Company)

Emerson, Willis George. (1965) *The Smokey God, or A Voyage to the Inner World*. (Mundelein, IL, Palmer Publishing) Originally published in 1908.

Fitting, Peter, (ed.). (2004) *Subterranean Worlds: A Critical Anthology*. (Middletown, CT, Wesleyan University Press)

Flanagan, G. Pat. (1973) *Pyramid Power*. (Camarillo, CA, De Vorss)

Frissell, Bob. (2002) *Nothing in This Book is True, But It's Exactly How Things Are: Third Edition, Revised and Expanded*. (Berkeley, CA, Frog, Ltd.)

Gardner, Marshall B. (1920) *A Journey to the Earth's Interior*. (Aurora, IL, Marshall B. Gardner) Originally published in 1913.

Gardner, Martin. (1957) *Fads and Fallacies in the Name of Science*. (New York, Dover)

Gerhard, Ken. (2007) *Big Bird! Modern Sightings of Flying Monsters*. (North Devon, CFZ Press)

Godfrey, Linda S. (2003) *The Beast of Bray Road: Tailing Wisconsin's Werewolf*. (Black Earth, WI, Prairie Oak Press)

—— (2006) *Hunting the American Werewolf: Beast Me in Wisconsin and Beyond*. (Madison, WI, Trails Books)

Godwin, Joscelyn. (1996) *Arktos: The Polar Myth in Science, Symbolism, and Nazi Survival*. (Kempton, IL, Adventures Unlimited Press)

Green, John. (1980) *On the Track of the Sasquatch*. (Blaine, WA, Hancock House)

Heron, Patrick. (2004) *The Nephilim and the Pyramid of the Apocalypse*. (New York, Kensington)

Heuvelmans, Bernard. (1972) *On the Track of Unknown Animals*. (Boston, M.I.T. Press)

Hunter, Don with René Dahinden. (1993) *Sasquatch/Bigfoot: The Search for America's Incredible Creature*. (Buffalo, NY, Firefly Books) Originally published in 1973.

Jessup, Morris K. (1955) *The Case for the UFO*. (New York, Citadel Press)

Joseph, Frank. (2004) *Survivors of Atlantis: Their Impact on World Culture*. (Rochester, VT, Bear and Company)

Joseph, Lawrence E. (2007) *Apocalypse 2012: A Scientific Investigation into Civilization's End*. (New York, Morgan Road Books)

Josyer, G. R., (trans.). (1973) *The Vaimanika Shastra*. (Mysore-4, India, Coronation Press)

Keel, John A. (1991) *The Mothman Prophecies*. (New York, Tor) Originally published in 1975.

—— (2002) *The Complete Guide to Mysterious Beings*. (New York, Tor) Originally published in 1970 as *Strange Creatures from Time and Space*.

Kelleher, Colm A. and George Knapp. (2005) *Hunt for the Skinwalker: Science Confronts the Unexplained at a Remote Ranch in Utah*. (New York, Paraview)

King, Godfre Ray [Guy Warren Ballard]. (1934) *Unveiled Mysteries*. (Chicago, St. Germaine)

Klement, Jan. (1976) *The Creature: Personal Experiences with Bigfoot*. (Elgin, PA, Allegheny Press)

Krantz, Grover S. (1999) *Bigfoot Sasquatch Evidence*. (Blaine, WA, Hancock House) Originally published in 1992.

Lasperitis, Jack 'Kewaunee.' (1998) *The Psychic Sasquatch and their UFO Connection*. (Mill Spring, NC, Wild Flower Press)

Leichtman, Robert R. (1980) *From Heaven to Earth: Nikola Tesla Returns*. (Atlanta, GA, Enthea Press)

Le Plongeon, Augustus. (1997) *Maya/Atlantis: Queen Moo and the Egyptian Sphinx*. (Whitefish, MT, Kessinger Publishing) Originally published in 1896.

Le Poer Trench, Brensley. (2005) *Finding Lost Atlantis Inside the Hollow Earth*. (Global Communications) Originally published in 1975 as *Secret of the Ages*.

Leslie, Desmond and George Adamski. (1953) *Flying Saucers Have Landed*. (New York, The British Book Centre)

Long, Greg. (2004) *The Making of Bigfoot: The Inside Story* (New York, Prometheus Books.)

Lyne, William R. (1998) *Occult Ether Physics: Tesla's Hidden Space Propulsion System and the Conspiracy to Conceal It* (Lamy, NM, Creatopia Productions).

MacGregor, Rob and Bruce Gernon. (2005) *The Fog: A Never Before Published Theory of the Bermuda Triangle*. (Woodbury, MN, Llewellyn Publications)

Maclellan, Alec. (1982) *The Lost World of Agharta: The Mystery of Vril Power*. (London, Souvenir Press)

—— (1992) *The Hollow Earth Enigma*. (London, Souvenir Press)

Manning, Jeane and Nick Begich. (2002) *Angels Don't Play this HAARP: Advances in Tesla Technology*. (Anchorage, AK, Earthpulse Press)

Meldrum, Jeff. (2006) *Sasquatch: Legend Meets Science*. (New York, Tom Doherty Associates)

Mott, William Michael. (2007) *Caverns, Cauldrons and Concealed Creatures*, 2nd edition. (Frankston, TX, T.G.S. Publishers)

—— (ed.). (2007) *This Tragic Earth: The Art and World of Richard Sharpe Shaver*. (Frankston, TX, T.G.S. Publishers)

Nichols, Preston B. (1992) *The Montauk Project: Experiments in Time*. (New York, Sky Books)

Norman, Eric. (1972) *This Hollow Earth*. (New York, Lancer Books)

Ocean, Joan. (1989) *Dolphin Connection: Interdimensional Ways of Living*. (Captain Cook, Hawaii, Dolphin Connection Press)

O'Donnell, Phillip. (2006) *Dinosaurs: Dead or Alive?* (Longwood, FL, Xulon Press)

Oliver, Frederick Spencer. (1905) *A Dweller on Two Planets*. (Los Angeles, Borden)

Oudemans, A.C. (2007) *The Great Sea Serpent*. (Landisville, PA, Coachwhip Press) Originally published in 1892.

Pauwels, Louis and Jacques Bergier. (1964) *The Morning of the Magicians*. (New York, Stein and Day)

Powell, Thom. (2003) *The Locals: A Contemporary Investigation of the Bigfoot/Sasquatch Phenomenon*. (Blaine, WA, Hancock House)

Pyle, Robert Michael. (1995) *Where Bigfoot Walks: Crossing the Dark Divide*. (Boston, Houghton Mifflin)

Randles, Jenny. (2002) *Time Storms*. (New York, Berkley Books)

Reed, William. (1906) *The Phantom of the Poles*. (New York, Walter S. Rockey Company)

Robbins, Dianne. (2003) *Messages from the Hollow Earth*. (Rochester, NY, Dianne Robbins)

Sanderson, Ivan T. (2006) *Abominable Snowmen: Legend Come to Life.* (Kempton, IL, Adventures Unlimited Press) Originally published in 1961.
Scott-Elliot, W. (1925) *The Story of Atlantis and the Lost Lemuria.* (London, Theosophical Publishing House) Originally published as two books, in 1896 and 1904.
Sheppard-Wolford, Sali. (2006) *Valley of the Skookum: Four Years of Encounters with Bigfoot.* (Enumclaw, WA, Pine Winds Press)
Smith, Warren. (1976) *The Hidden Secrets of the Hollow Earth.* (New York, Kensington Publishing)
Standish, David. (2006) *Hollow Earth: The Long and Curious History of Imagining Strange Lands, Fantastical Creatures, Advanced Civilizations, and Marvelous Machines Below the Earth's Surface.* (New York, Da Capo)
Storm, Margaret. (1959) *Return of the Dove.* (Baltimore, Margaret Storm)
Swartz, Tim. (no date) *The Lost Journals of Nikola Tesla: HAARP – Chemtrails and the Secret of Alternative 4.* (New Brunswick, NJ, Global Communications)
—— (ed.). (2005) *Richard Shaver: Reality of the Inner Earth.* (New Brunswick, NJ, Global Communications)
Teed, Cyrus Reed. (1922) *The Cellular Cosmogony.* (Estero, FL, The Guiding Star Publishing House) Originally published in 1898.
Tesla, Nikola and David H. Childress. (1993) *The Fantastic Inventions of Nikola Tesla.* (Steele, IL, Adventures Unlimited Press)
Toth, Max. (1985) *Pyramid Power.* (Rochester, VT, Destiny Books)
Wauchope, Robert. (1962) *Lost Tribes and Sunken Continents: Myth and Method in the Study of American Indians.* (Chicago, University of Chicago Press)
Whitcomb, Jonathan. (2006) *Searching for Ropens: Living Pterosaurs in Papua New Guinea.* (Livermore, CA, BookShelf Press)
X, Commander. (no date) *The Philadelphia Experiment Chronicles: Exploring the Strange Case of Alfred Bielek and Dr. M.K. Jessup.* (New Brunswick, NJ, Abelard Productions, Inc)
X, Michael. (1992) *Tesla: Man of Mystery.* (New Brunswick, NJ, Inner Light Publications)

Index